Lake Springfield
in Illinois

Lake Springfield in Illinois

Public Works and Community Design in the Mid-Twentieth Century

ROBERT MAZRIM AND CURTIS MANN

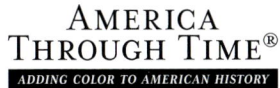

About the Authors

ROBERT MAZRIM has worked in Midwestern archaeology and history for thirty years. He currently serves as Associate Scientist Historic Resources Archaeologist at the Illinois State Archaeological Survey, and is the Executive Director of the Foundation for Colonial and American Studies. Mazrim is the author of six books and numerous journal articles concerning the archaeology and history of the Midwest.

CURTIS MANN is the manager of the Sangamon Valley Collection at Lincoln Library, the public library of Springfield, Illinois. Mann graduated from Southern Illinois University with a B.A. in History and the University of Illinois with a M.S. in Library and Information Science. He has co-authored ten pictorial books about the history of Springfield.

America Through Time is an imprint of Fonthill Media LLC
www.through-time.com
office@through-time.com

Published by Arcadia Publishing by arrangement with Fonthill Media LLC
For all general information, please contact Arcadia Publishing:
Telephone: 843-853-2070
Fax: 843-853-0044
E-mail: sales@arcadiapublishing.com
For customer service and orders:
Toll-Free 1-888-313-2665

www.arcadiapublishing.com

First published 2021

Copyright © Robert Mazrim and Curtis Mann 2021

ISBN 978-1-63499-293-0

All rights reserved. No part of this publication may be reproduced, stored in a retrieval system or transmitted in any form or by any means, electronic, mechanical, photocopying, recording or otherwise, without prior permission in writing from Fonthill Media LLC

Typeset in 10pt on 13pt Sabon
Printed and bound in England

Contents

1	The Need for Lakes	7
2	Before the Lake	9
3	Building a Lake	18
4	New Neighborhoods on the Water	43
5	Things to Do	54
6	At the Clubs	74
7	The Drought of 1952-55	86
8	Life on the Lake	92

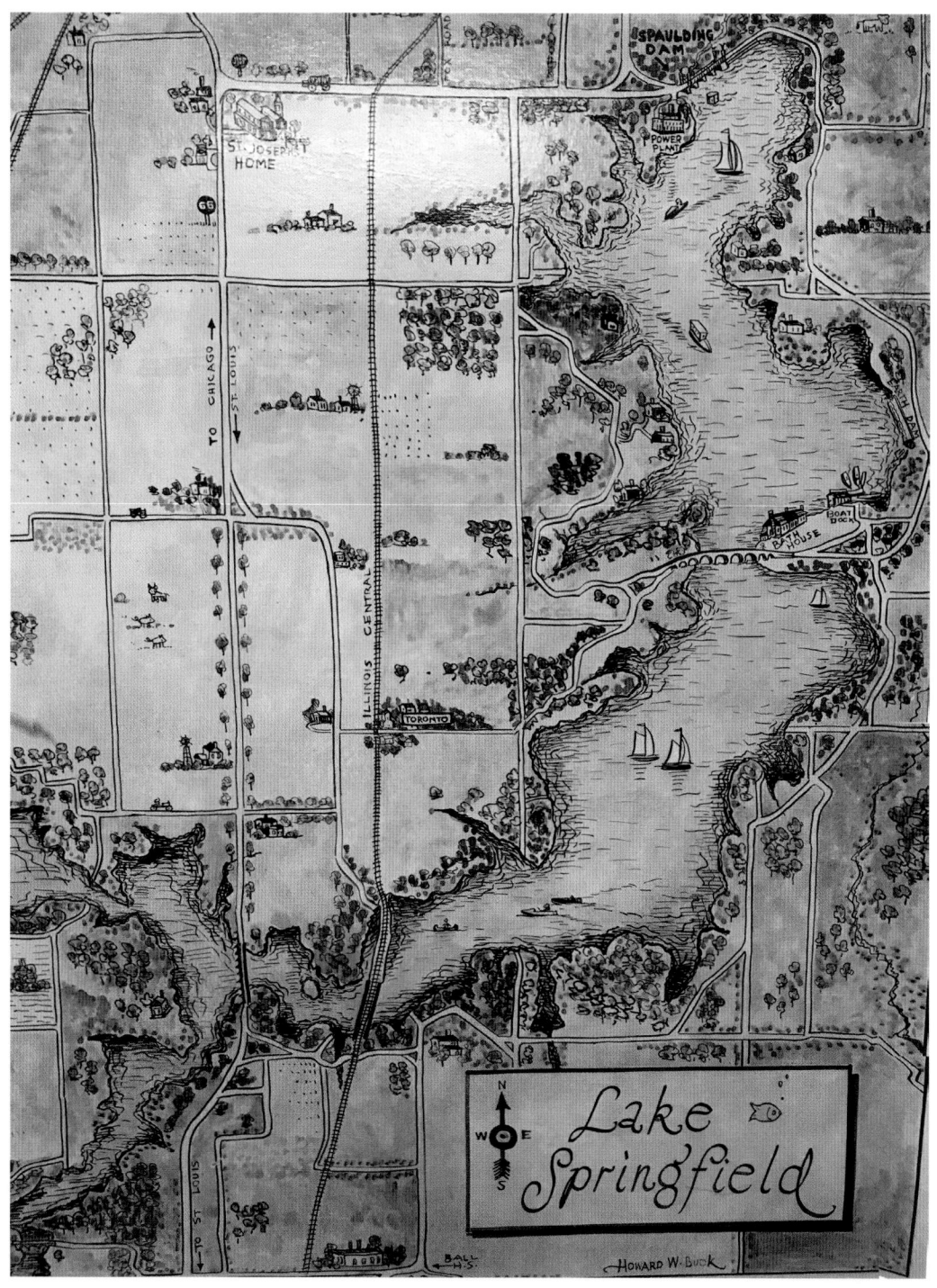

Folk-painted map of Lake Springfield by Howard Buck, *circa* 1937. [*Sangamon Valley Collection, Lincoln Library*]

1

The Need for Lakes

People need water, and Illinois is certainly a well-watered state. Its lands are crossed by the Mississippi, Illinois, Ohio, and Wabash Rivers and their many tributaries, as well as the adjacent Lake Michigan at Chicago. However, natural lakes are reasonably few in the state, and are located primarily in the extreme northern and southern portions of the state.

After World War I, growing urban communities located in the once prairie-covered uplands needed more reliable sources of water. Before the war, towns and cities generally relied on individual wells, cisterns, and municipal water towers and reservoirs.

After the war, man-made lakes created by the damning of local creeks provided consistent sources of large water supplies. Another important function of artificial lakes was to provide water for new coal-fired, steam-generated power plants. Of course, those power plants also represented the growing needs of expanding communities across the state, including the city of Springfield.

Official consideration of a municipal lake to serve the needs of Springfield began as early as 1925. Initially, a site on the Sangamon River north of the city was proposed. Soon afterward, however, an alternate location on Sugar Creek (south of the city) was chosen. Lake Springfield was initially designed as a water supply, but plans for a new coal-powered electrical power plant were incorporated into its design during construction. The new lake and power plant were the successors to the city's water works and power plant located north of the city along the Sangamon River. In addition, the lands immediately surrounding the lake became part of a planned residential and recreational community. As early as 1925, city planners envisioned the potential benefit not only of water and power, but of "facilities for water sports and lake-side cottage sites."

Lake Springfield was one of the earliest man-made lakes in the state, and is also among the largest. The lake itself covers over 4,000 acres of former floodplain affiliated with Sugar and Lick Creeks in Sangamon County. An additional 3,500 acres were purchased for shoreline residential development, park land, and municipal facilities.

The City of Springfield began purchasing land for a new lake in 1931, and construction of a dam over Sugar Creek began in 1933. The creek valleys were cleared of trees, brush, fence lines, houses, and agricultural buildings such as barns and sheds. Rains came, and by the spring of 1935, the shallow creek valleys had become a 15-mile long lake, located on the southern outskirts of the city of Springfield. The construction of the lake, including the purchase of land, its clearing, and the construction of a dam and new roads, cost approximately 2.5-million dollars. Today, the lake provides water and power for nearly 150,000 residents of central Illinois.

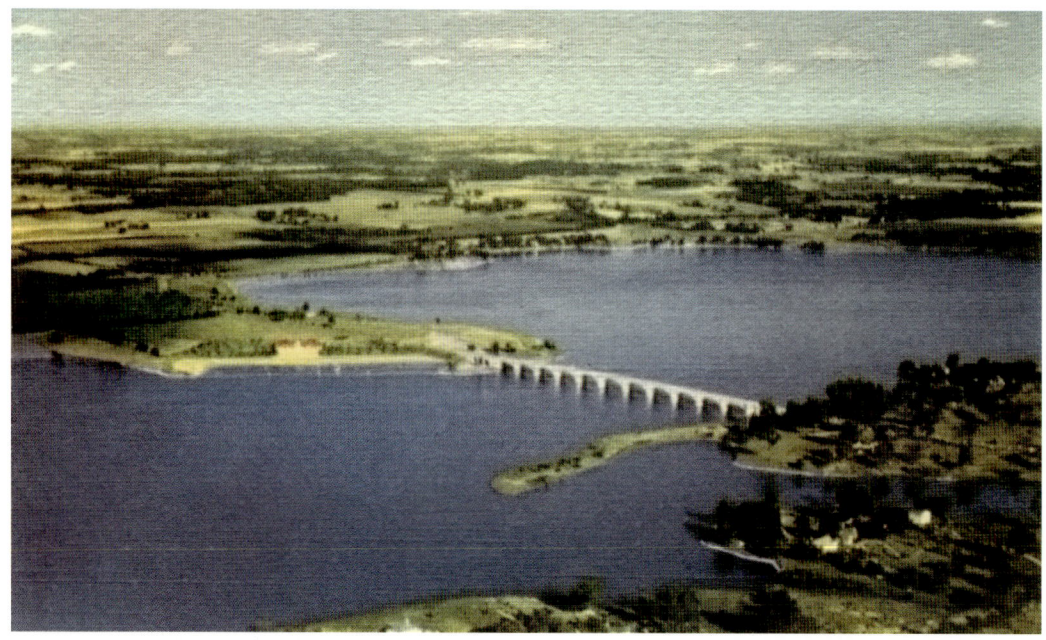

Stylized postcard view of Lindsay Bridge spanning a narrows on Lake Springfield. [*Sangamon Valley Collection, Lincoln Library*]

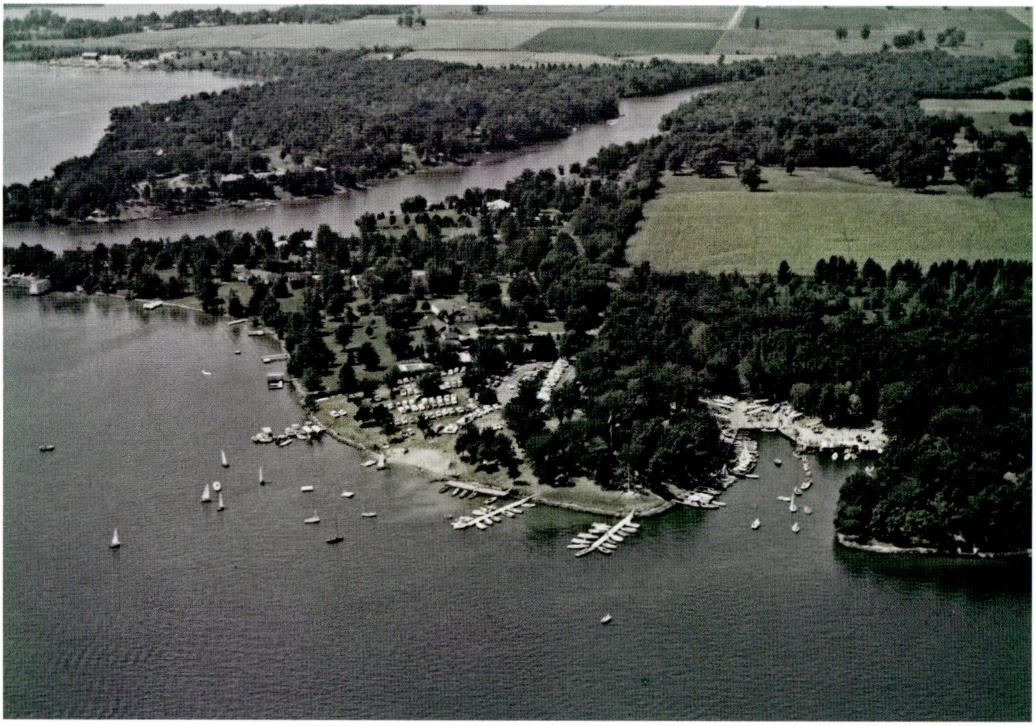

Aerial view of the Island Bay Yacht Club, *circa* 1960. [*Sangamon Valley Collection, Lincoln Library*]

2

Before the Lake

Lake Springfield was constructed by the damming and flooding of portions of Sugar and Lick Creek, small tributaries of the Sangamon River. Like most creek valleys in the Midwest, the cultural history in and around the Sugar Creek Valley is a rich and deep one. Evidence of prehistoric societies that once hunted and camped in the valley is occasionally found on the shoreline in the form of ancient stone tools. Much older are fossil deposits on the southwestern shore of the lake, which include remains of ancient marine life (such as trilobites, cephalopods, crinoids, and ammonites) from a time when the entire region was an ocean floor.

European history along the bluffs that would become the shoreline of Lake Springfield began before Europeans settled in the area. Along the eastern timberline of Sugar Creek was an ancient trail followed by French colonial settlers who were living in villages 100 miles to the south. French records indicate that the old trail was used as early as 1711, when a French priest made a journey across central Illinois with a Native American hunting party. Archaeological evidence suggests the trail might actually date well into prehistory.

Euro-Americans began to settle the Springfield area shortly after the close of the War of 1812. The first settlement of the portion of the valley that would become Lake Springfield dates to the late 1810s. The small community of Cotton Hill was settled in the 1820s, and was centered on a saw and gristmill known as Crow's Mill. It was located directly along Sugar Creek and included a saw and grist mill, a general store, a blacksmith shop, and several houses. Cotton Hill was abandoned, dismantled, and flooded as part of the construction of the lake.

Nearby, a small limestone quarry at the base of the creek valley supplied stone for the impressive Greek Revival capitol building in Springfield, built in 1837. Some unfinished stones intended for the columns of the capitol building were found abandoned in the old quarry during the construction of lake. Some of these were dragged up to the bluff top and can still be seen today.

Also flooded over in the 1930s was a small segment of the legendary Route 66 corridor that passed through Cotton Hill as it headed north to the city of Springfield. Numerous family farms were also in the path of the new reservoir. In fact, a few of the houses on the shores of the twentieth-century lake are actually much older, once serving as farmhouses on the bluffs overlooking the creek valley. During the early 1930s, some local historians suggested the name "Sugar Lake" for the new reservoir, as a reference to the creek itself and to the rich history of Euro-American settlement there.

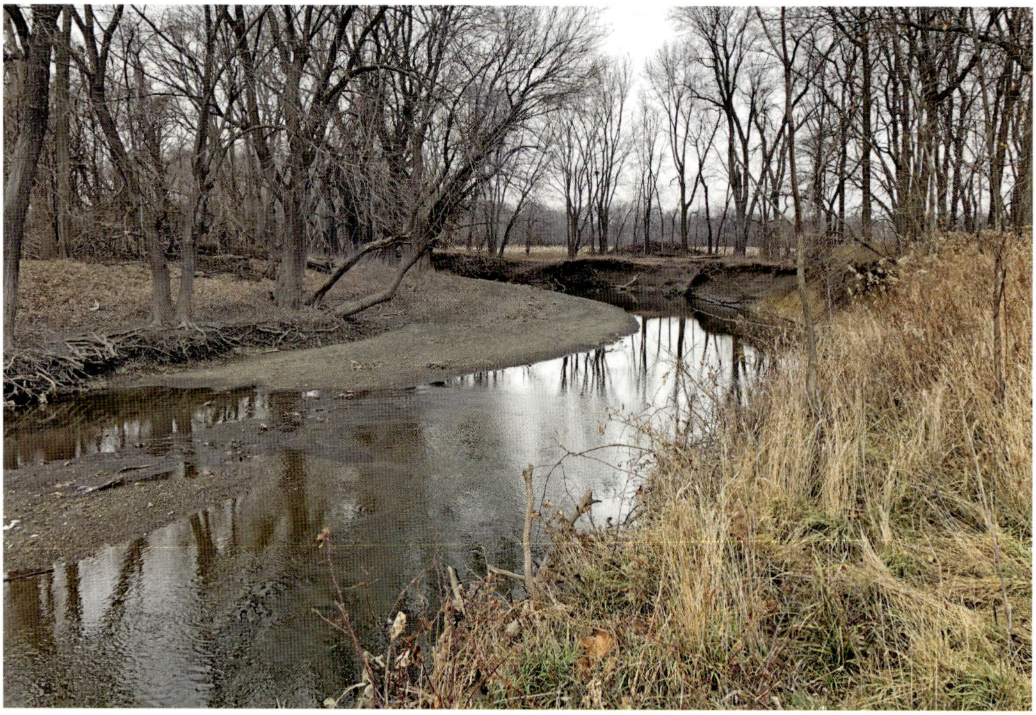

Lake Springfield in Illinois

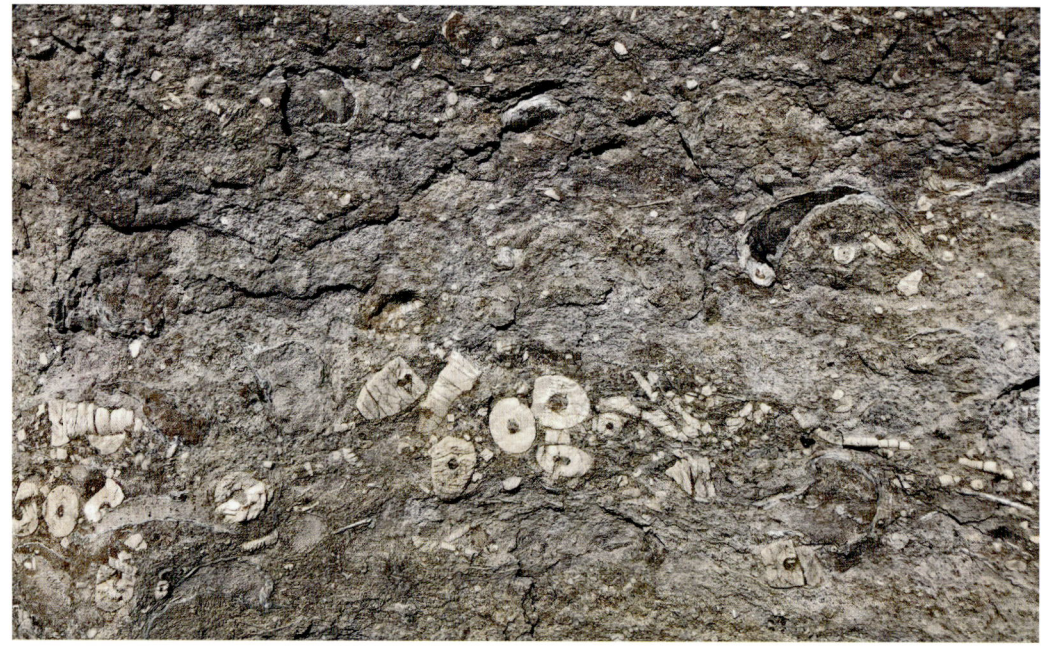

Lake Springfield is an artificial reservoir created by the damming of Sugar Creek (above), a tributary of the Sangamon River. The limestone bedrock of Sugar Creek reveals a much more ancient watery past—fossils of marine life from over 300 million years ago, when what would become Illinois was part of a shallow tropical ocean. [*Courtesy of Robert Mazrim*]

Before the Lake

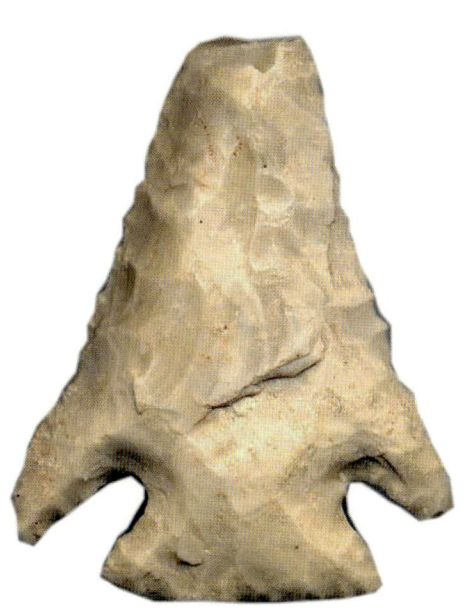
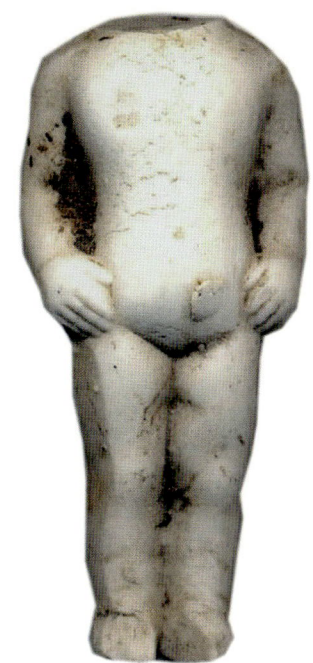

People lived in the Sugar Creek Valley long before the first homes were built along the new shore of the new lake. Over 10,000 years prior to the arrival of the first Euro-Americans, the Sugar Creek Valley was home to a number of Native American societies. The stone spear point on the left is approximately 8000 years old. The porcelain doll on the right dates to the late nineteenth century. They were found within five feet of each other on the shore of the lake. Also shown here is an ancient "nutting stone," found during low water. [*Courtesy of Robert Mazrim*]

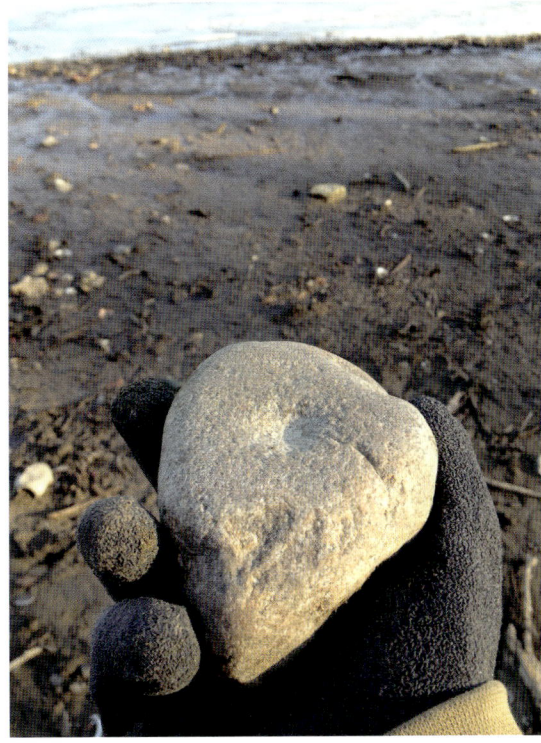

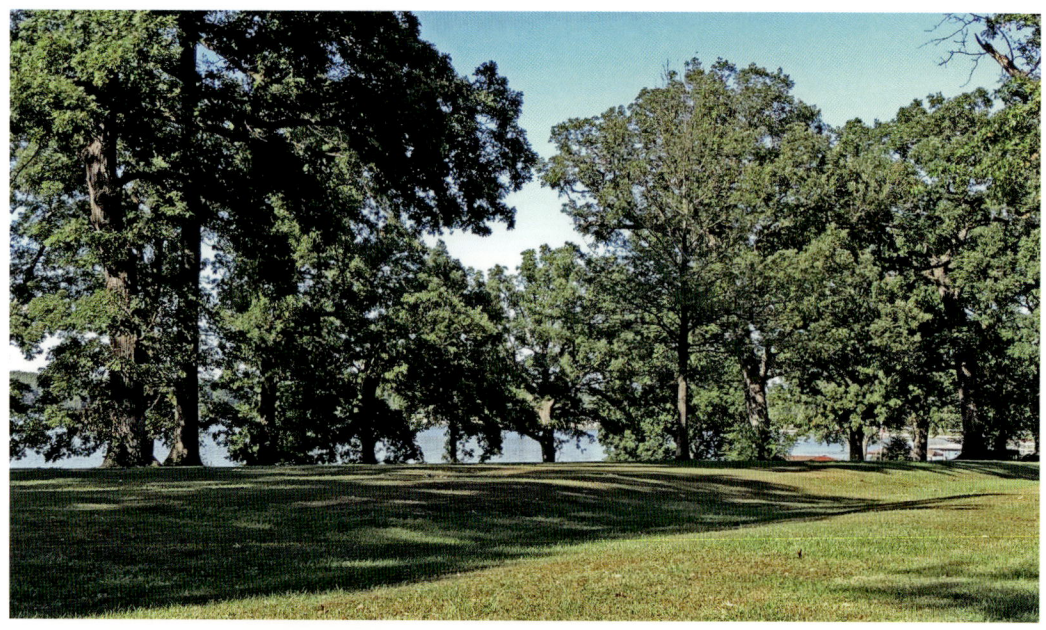

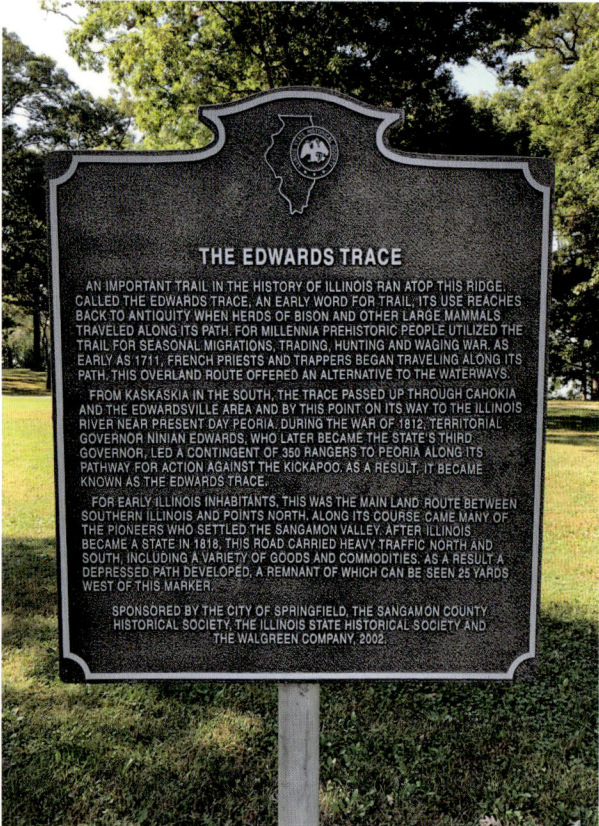

Still visible in Central Park on the eastern shore of the lake is a deep rut left behind by an ancient overland trail that crossed through the area. The trail was used by the French as early as the early 1700s, and may actually be prehistoric in origin. The trail is known as "Edwards' Trace" after its use by Territorial Governor Edwards during the War of 1812. [*Courtesy of Robert Mazrim*]

Before the Lake

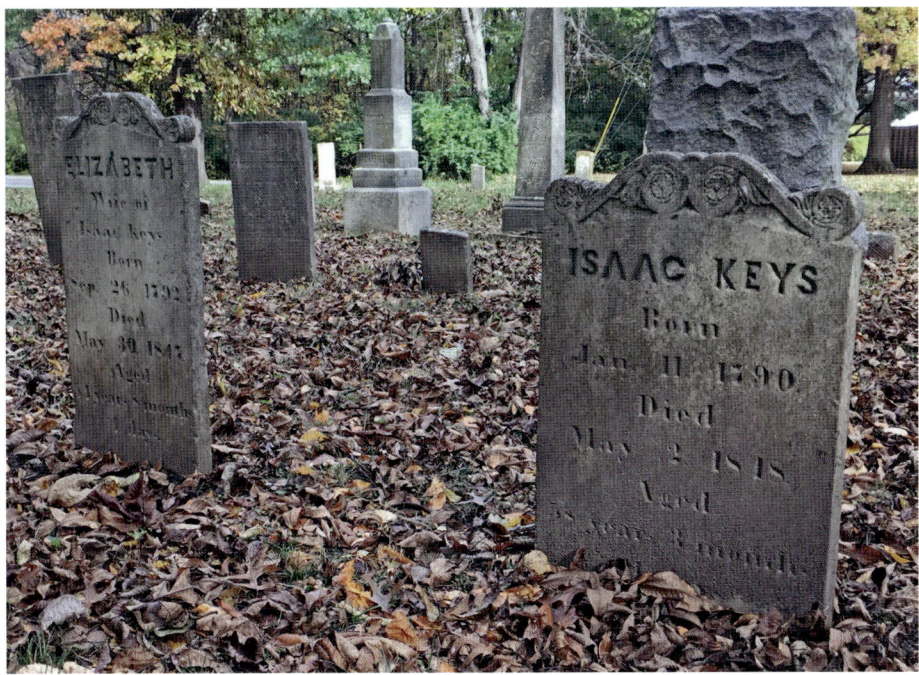

Euro-Americans begin to settle the portion of the Sugar Creek Valley that would become Lake Springfield in 1818. Early family cemeteries still dot the shoreline of the lake. Isaac Keys, whose gravestone is shown above, was one of the earliest settlers of the area. David Brunk (below) arrived in the 1820s as a young man, and made wheel thrown crockery at the edge of the Sugar Creek timberline until his death in 1855. [*Courtesy of Robert Mazrim*]

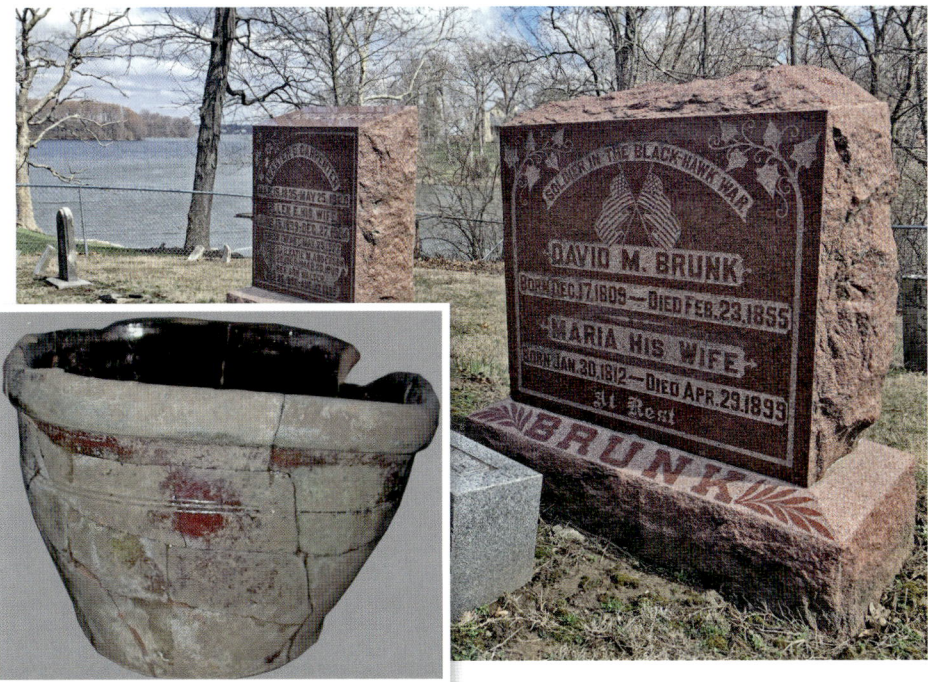

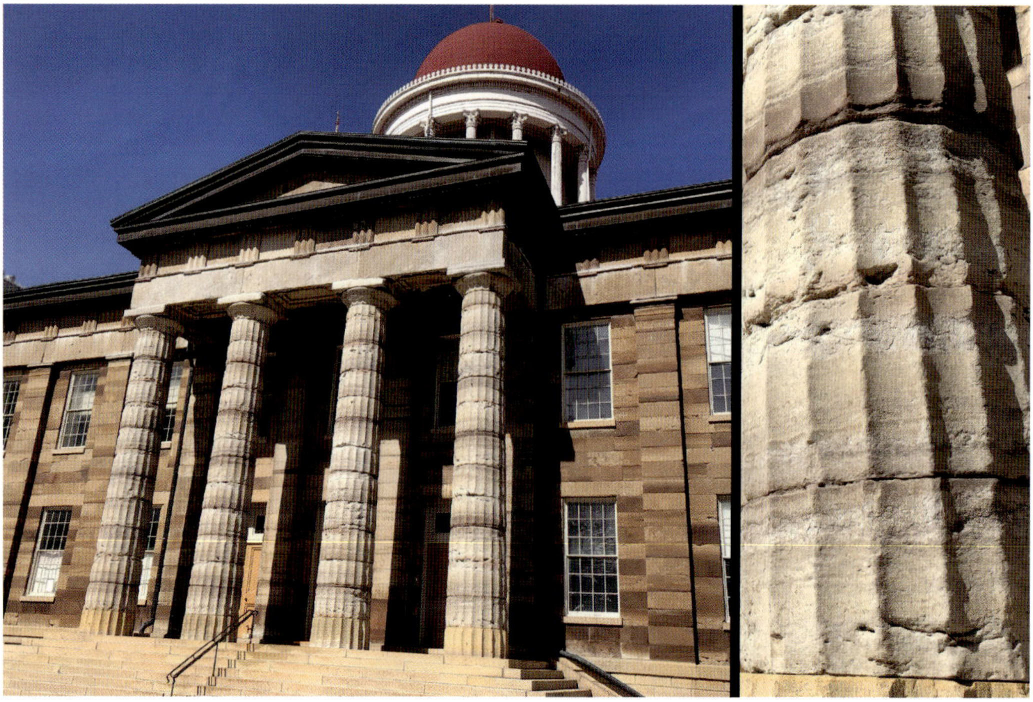

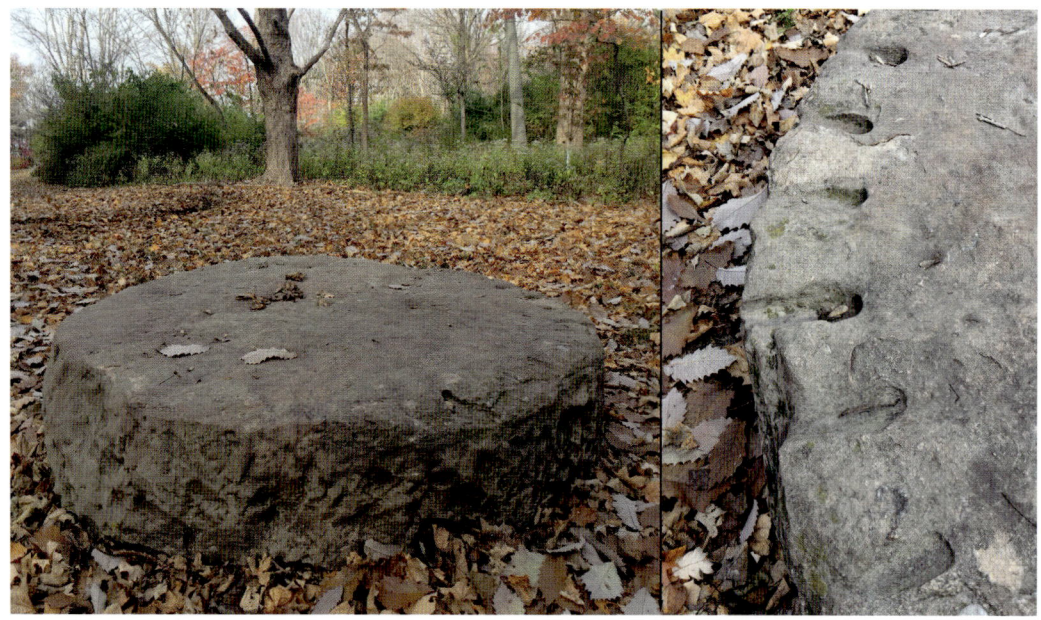

Now under water are the remains of a small limestone quarry that was used during the 1830s. Most notably, stone from the quarry was utilized in the construction of the impressive Greek Revival State Capitol building in nearby Springfield, constructed in 1837. Still preserved in two parks along the lake are leftover disc-shaped "blanks" intended for use in the column facade of the Capitol building. [*Courtesy of Robert Mazrim*]

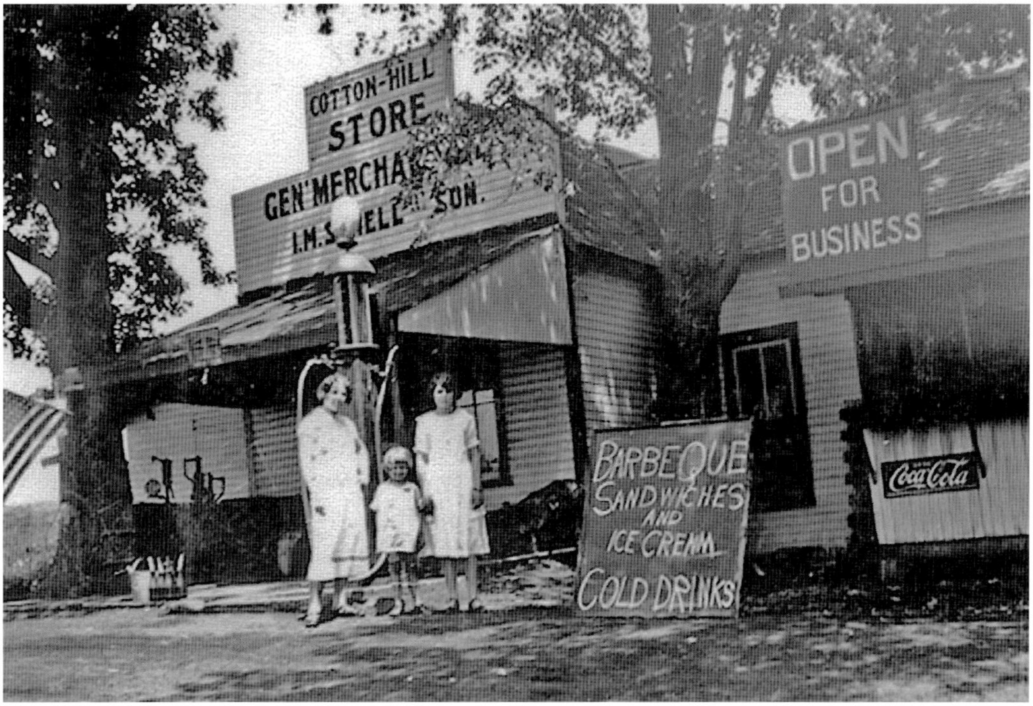

Around 1830, a small village and mill was established along Sugar Creek. The village was called "Cotton Hill" and was about 100 years old when it was vacated and dismantled to make way for the new reservoir. Just before its abandonment, Cotton Hill included an abandoned grist mill ("Crow's Mill"), a small grocery store, a blacksmith shop, a gas station, a school, and several homes. [*Sangamon Valley Collection, Lincoln Library*]

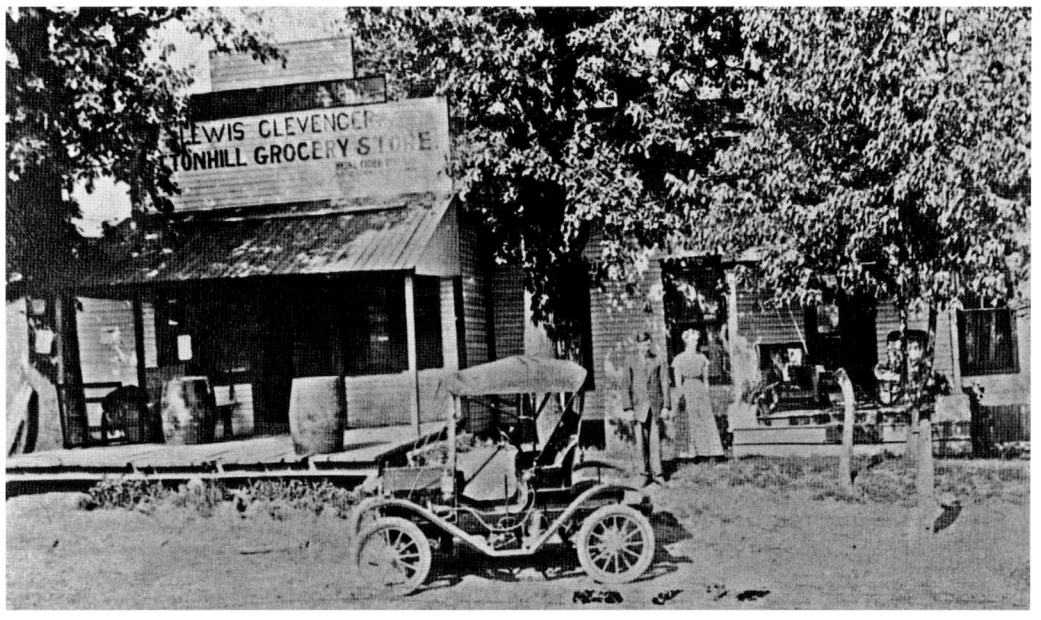

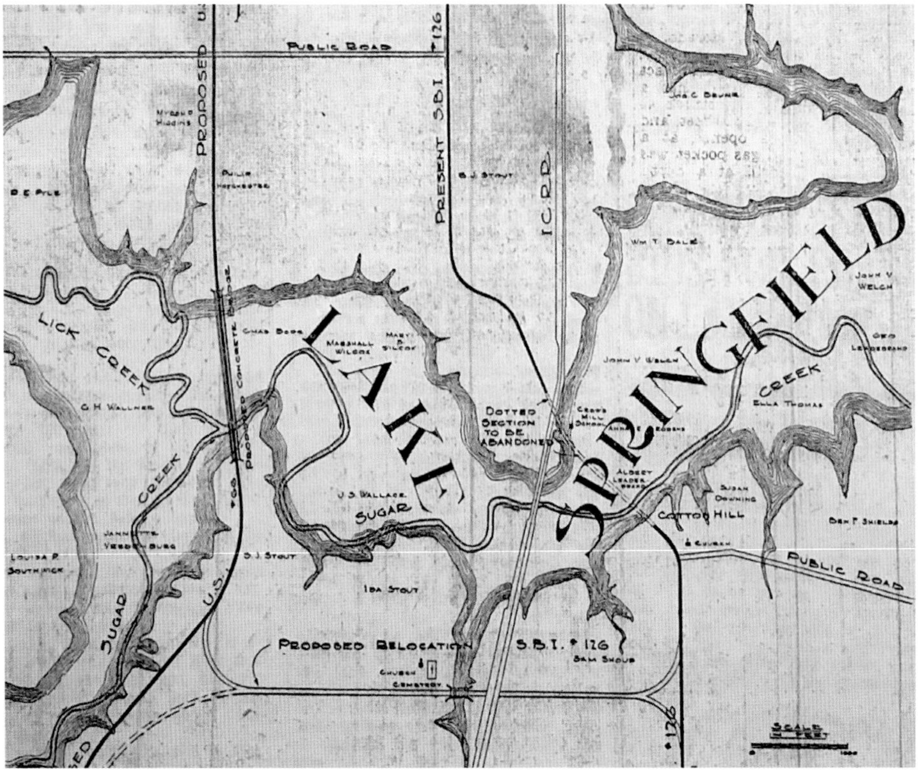

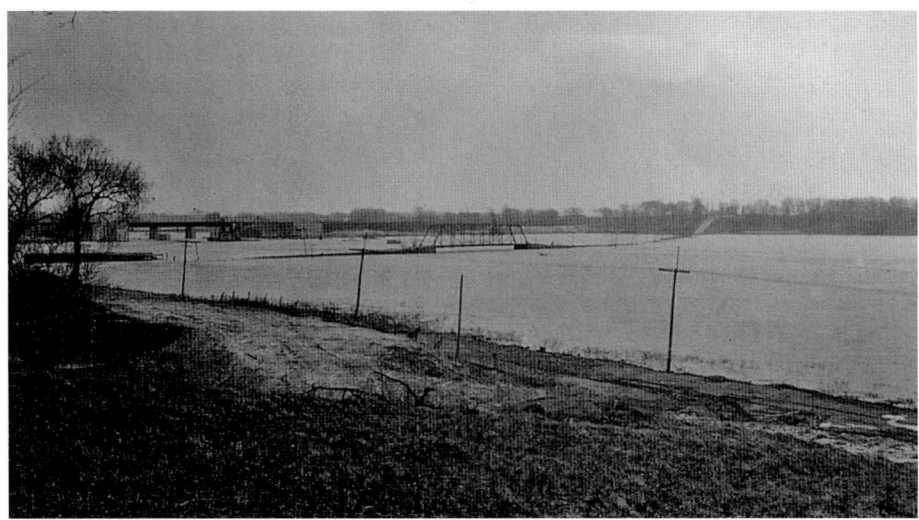

In this 1930s map showing the relocation of roads affiliated with the construction of the lake, the actual site of Cotton Hill is depicted by the small square near the "R" in "Springfield," which marks the location of the village school. The road that passed through the village was an early leg of Route 66, shown below in a photograph of a flood that pre-dated the construction of the lake. That road was relocated to the west (to become Route 66 and later Interstate 55) and the village was abandoned and demolished. [*Sangamon Valley Collection, Lincoln Library*]

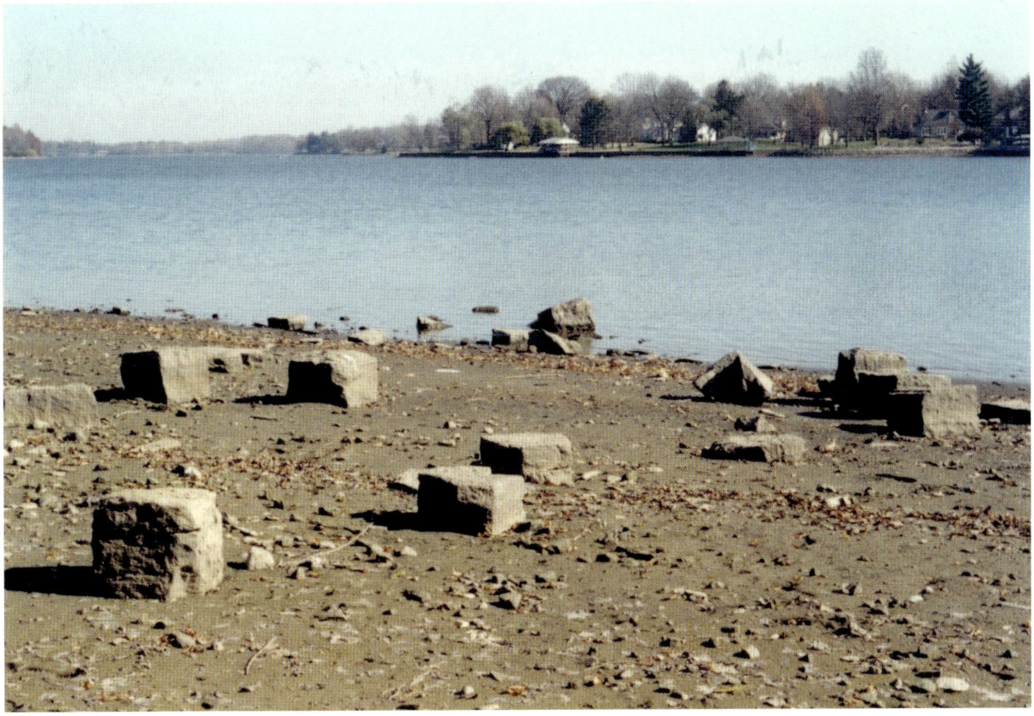

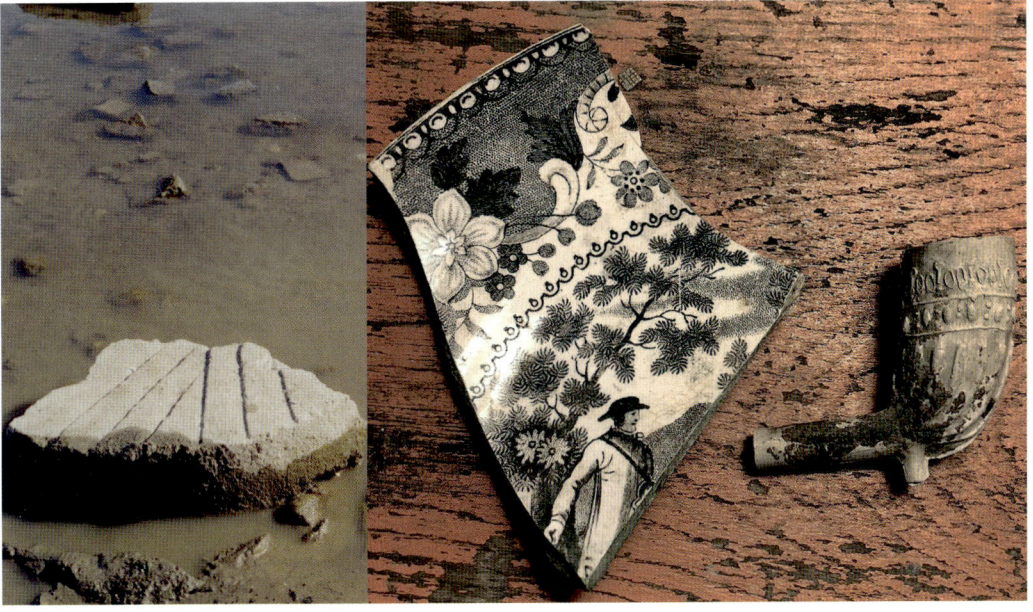

Occasionally, in times of drought, low water reveals the archaeological remains of the village of Cotton Hill. Shown above are crude limestone piers that may be affiliated with Crow's Mill. Below, a broken millstone lies in the mud, near a variety of artifacts associated with frontier life in that village, including this 1830s Staffordshire saucer fragment and early British smoking pipe. [*Courtesy of Robert Mazrim*]

3

Building a Lake

Plans for a reservoir water supply for the growing city of Springfield were underway by the early 1920s, in part as a response to a severe drought in 1914. At the time, the city drew most of its water from the Sangamon River, north of town. Five years of studies and surveys followed, resulting in several recommended sites. It soon became clear that the damming of a portion of Sugar Creek just south of the city was the most pragmatic choice. Plans for the reservoir on the Sugar Creek valley were published in May of 1930.

The planning, construction, maintenance, and operation of the lake and its facilities has from the beginning been overseen by City Water Light and Power (CWLP), a municipally owned utility that manages the power plant, the water treatment plant, the reservoir, and the surrounding property. Willis J. Spaulding, who served as superintendent of the city water works until 1911 when he was elected commissioner of public property, oversaw the construction of the lake as well as the new power plant built on its shores. His brother, Charles H. Spaulding, was the general superintendent of the Springfield Water Department and invented the Spaulding precipitator. Together, the Spaulding brothers shaped much of the design and construction of the reservoir and its infrastructure that has been critical to the habitation of the city of Springfield and surrounding communities for over eighty years. During the construction of the lake, many suggested the name "Lake Spaulding."

The construction of the new lake began in early 1931, beginning with the acquisition of 110 private properties, consisting mostly of farms on the floodplain, and the bluff line that would become the shore of the lake. The largest single purchase from a single family was from the S. J. Stout family, encompassing 338 acres of their family farm. Property owners were given market value prices for their land, but the municipal project was supported by eminent domain, and thus landowners had no real choice but to sell. Some were less than happy about being forced to leave their farms. In the fall of 1933, Leander Shoup held police at bay with a shotgun on his farm along the southern edge of Sugar Creek. After negotiations that began in the woods behind his house, it was agreed that he could, for a short time, continue to use his farmland that would ultimately be flooded by the lake. He was also issued a $9,755 check for his land. By 1935, however, he was living in a one-room apartment in downtown Springfield, and he died that year.

Clearing of trees and brush began in the fall of 1931. Crews of laborers used saws, heavy equipment, and even TNT to remove the vegetation on the floodplain that would soon be underwater. Some of the oak trees felled for the construction of the lake were over 200 years old. As recently as the early 1980s, massive tree stumps could still be seen during times of low water, blackened by decades underwater.

As construction progressed, fences were removed and small bridges across the creek were dismantled. Any structures that were to be inundated were demolished (which included what was left of the small village of Cotton Hill), as well as outbuildings, barns, and a few houses. One of those houses was situated on a slight hill on the floodplain, which after the lake was completed became an island. The nineteenth-century brick-lined cellar from that house is still visible in the woods today.

Construction of a dam across Sugar Creek began in the spring of 1933. The 1,700-foot-wide earthen dam was built with a concrete core. At the west end of the dam are five gates that are raised and lowered to control the lake level. The dam also serves as a roadway crossing the northern most edge of the lake, and is a popular fishing spot due to the deep water nearby. It was dedicated as "Spaulding Dam" in 1935, in honor of Willis Spaulding.

Several bridge projects were undertaken as part of the construction project. The largest were the Route 66 Bridge on the west side of the lake, and Lindsay Memorial Bridge on the east. The Route 66 Bridge (initially known as the South Sixth Street Bridge and now as the Interstate Highway 55 Bridge) represented a slight shift in the alignment of the Route 66 Highway that originally passed through the village of Cotton Hill to the east. The Lindsay Bridge (initially known as Fox Road Bridge) was so renamed for the poet Vachel Lindsay, who was born in Springfield. The bridge is of a somewhat rare concrete, cantilevered arch design based loosely on the Arroyo Seco Bridge in Pasadena, California. Vachel Lindsay's widow unveiled a bust of the writer at the western approach to the bridge during dedication ceremonies in 1935.

A 280-acre natural area on the western end of the lake provided for erosion control near the headwaters of the reservoir, but also preserved some of the natural habitat of the flooded valley and ultimately became a wildlife preserve. Small hillocks in the floodplain became islands after the valley was flooded. The three largest remain today as natural areas, while the smallest (on the extreme western end of the lake) disappeared due to erosion after 2000.

Once the dam was completed, the flooding of the valley took less than two years. Older residents on the surrounding bluff tops watched as the old fields gradually filled with water that would never recede. The lake reached full pool—at 560 feet above sea level—in May of 1935. A three-day dedication ceremony was held over the weekend of July 12, 1935. Boat races, choral concerts, performances of "Pinafore" and "Chimes of Normandy," and lakeside Vespers services were part of the ceremonies. On the new beach, a large crowd watched as "Neptune and Aphrodite rose from the water" at 7:30 p.m. on Saturday. Also, part of the festivities was the pouring of water from over thirty rivers, lakes, and oceans into the new lake. Water from the headwaters of the Mississippi, the Nile, the Red Sea, and the Thames were mixed into the water that flowed into the lake from Sugar and Lick Creeks.

In 1985, a state water survey reported that 6.5-million tons of sediment (from agricultural-related erosion surrounding Sugar and Lick Creeks to the west) had caused the lake's storage capacity to drop to 17-billion gallons, or approximately 13% less than its original holding capacity. Between 1987–1990, CWLP undertook a $7.8-million dredging project in the upper reaches of the lake to remove over 3-million cubic yards of sediment, reclaiming 650-million gallons of lost capacity.

Once the lake was constructed, municipal water and power facilities were constructed. Plans for the water treatment plant were unveiled in the fall of 1934. The plant initially included four filters and three clarifying basins, and featured new forms of water clarifiers that were designed by Charles Spaulding. His designs became models for later water treatment plants across the country. Water began flowing through the purification plant in October of 1936, and the Lakeside Power Plant was dedicated at the same time.

In 1942, the Illinois National Guard constructed a barracks near Spaulding Dam to house guardsmen who were stationed there to protect the dam, water, and power supply during World War II. After the war, the building was remodeled to serve as a naval reserve training center. Today, it is a property management center for CWLP.

A new coal-fired facility, the Dallman Power Station, was constructed next to the 1930s Lakeside Station in 1968. New boilers and generators were added in 1972 and 1978. The construction of the 200-megawatt Dallman Unit #4 station in 2009 replaced the 1960s units.

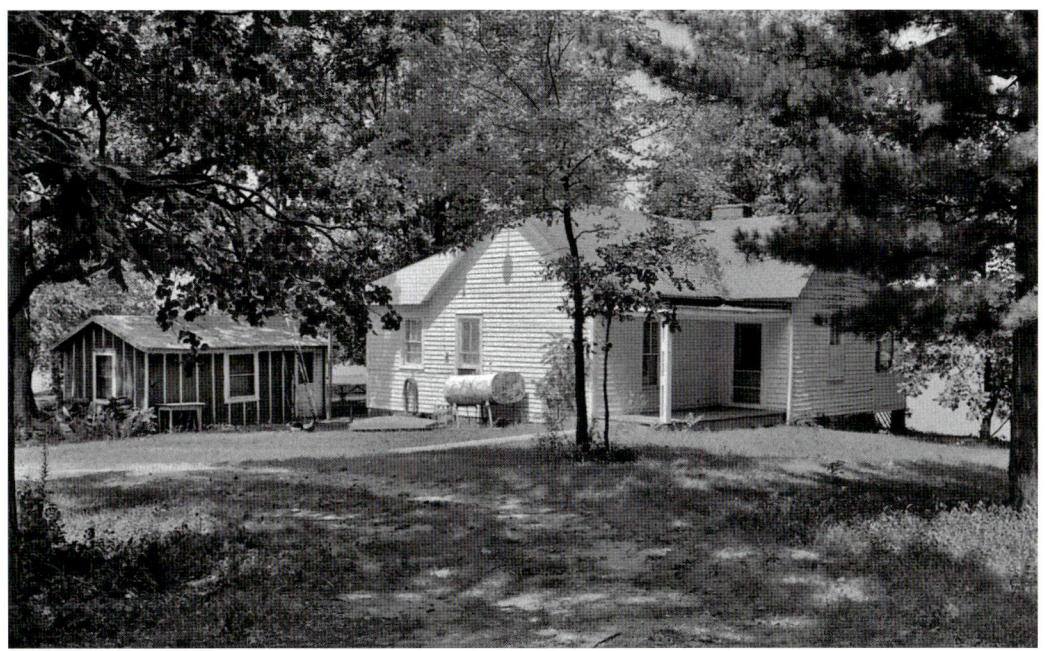

Over 100 properties were purchased for the construction of the lake. Most of these were farms. The small farmhouse above, built in the 1870s, became one of the first clubhouses on the lake during the mid-1930s. The farmhouse below was the home of Leander Shoup. Distraught by the forced buy-out of a portion of his farm, he held off authorities at gunpoint in the forests behind the house. The house, built before the Civil War, is now part of the Lincoln Memorial Gardens complex. [*Courtesy of Robert Mazrim*]

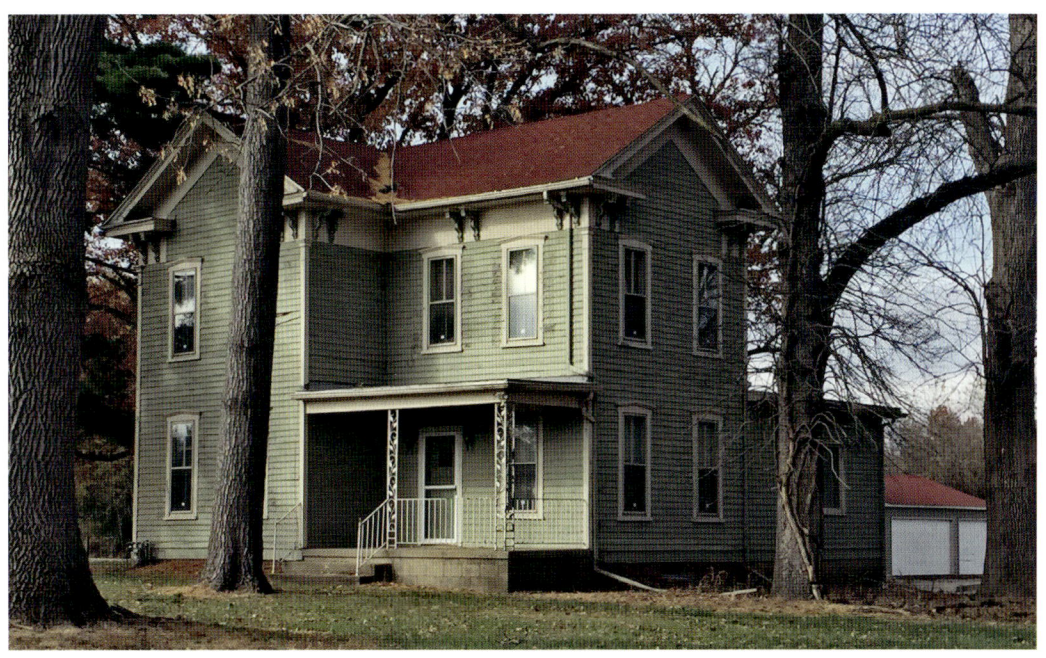

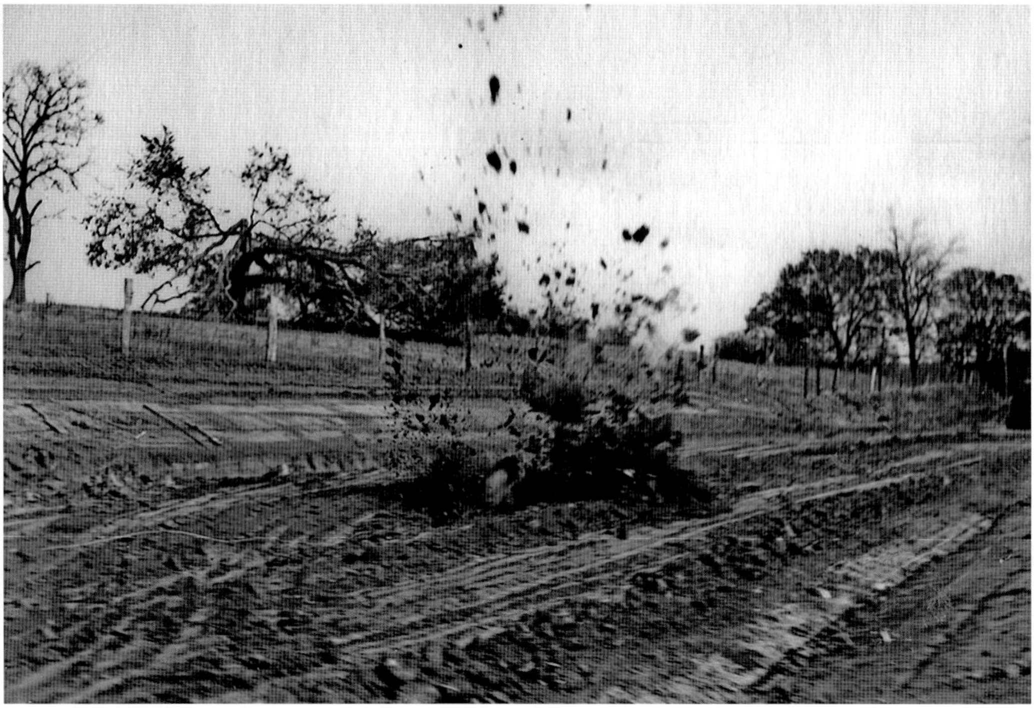

Clearing the land of obstructions such as small trees and underbrush was easy compared to tough tree stumps. Dynamite had to be employed to blast the stumps out of the ground, as shown above. An experienced crew was needed for such a delicate operation. [*Sangamon Valley Collection, Lincoln Library*]

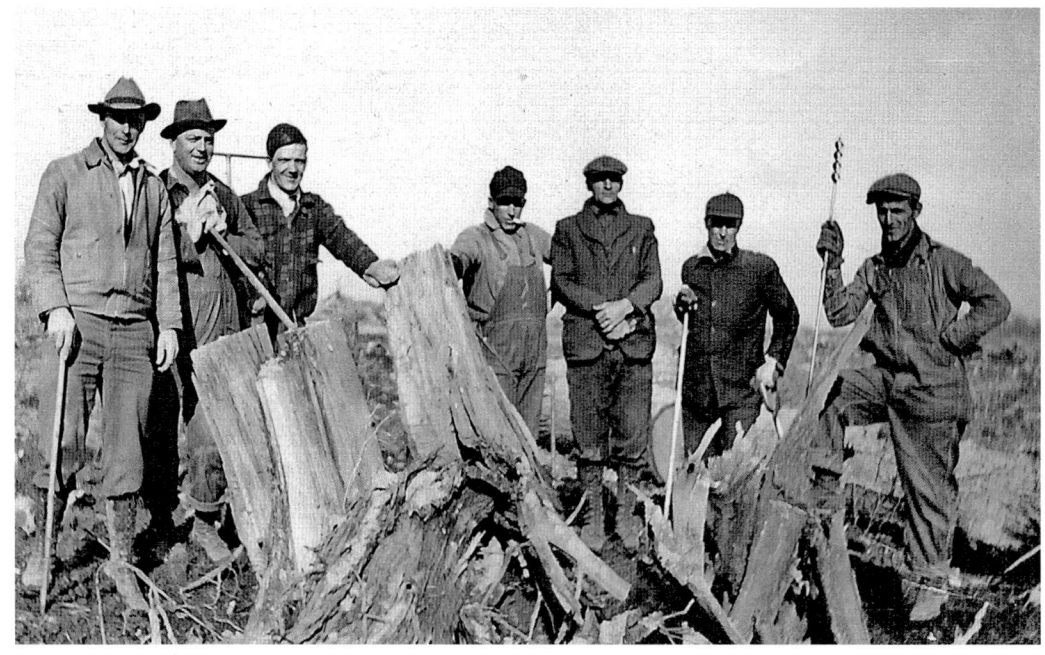

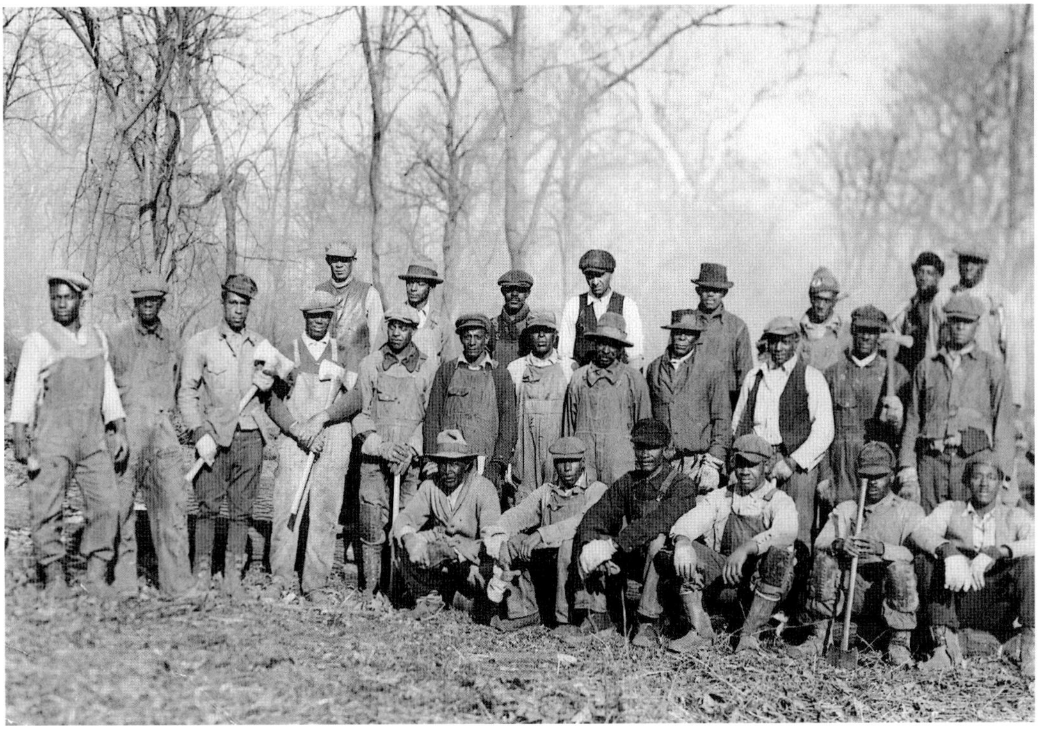

Logs suitable for sawing were saved to be later sold for lumber. Crews of laborers like the one shown above were paid fifty cents an hour to remove lumber from the lake site.

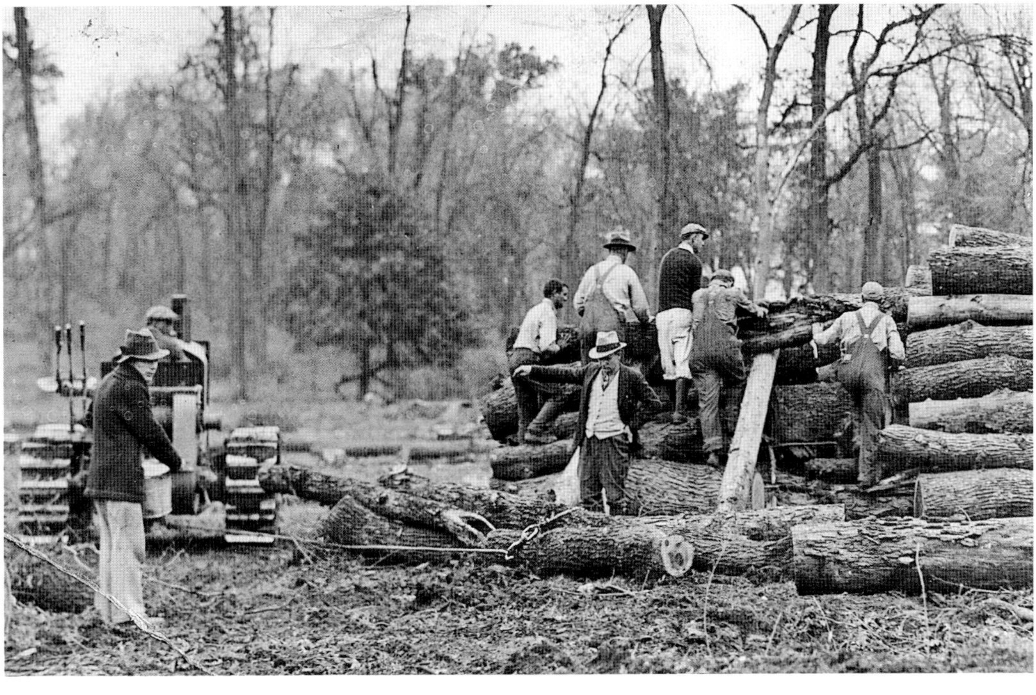

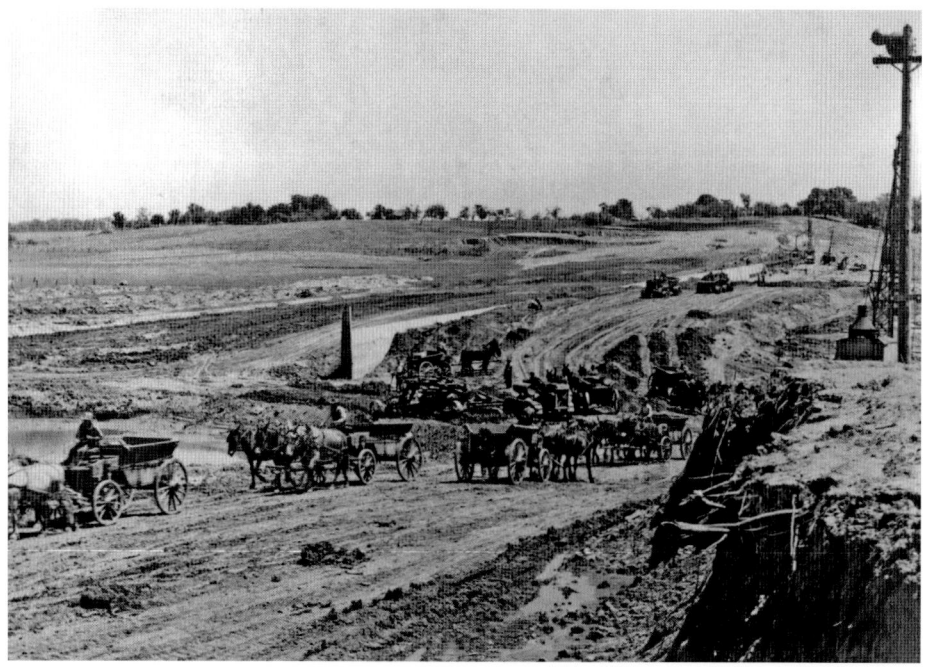

After the clearing of the timber, the construction of the dam at the north end of the impoundment was the next task. In the upper photograph, the base of the concrete wall can be seen, behind which is a large earthen support. Horse-drawn wagons were still being used to haul soil in the early 1930s. Below is a photograph of the concrete forms used to cast the core of the dam itself. [*Sangamon Valley Collection, Lincoln Library*]

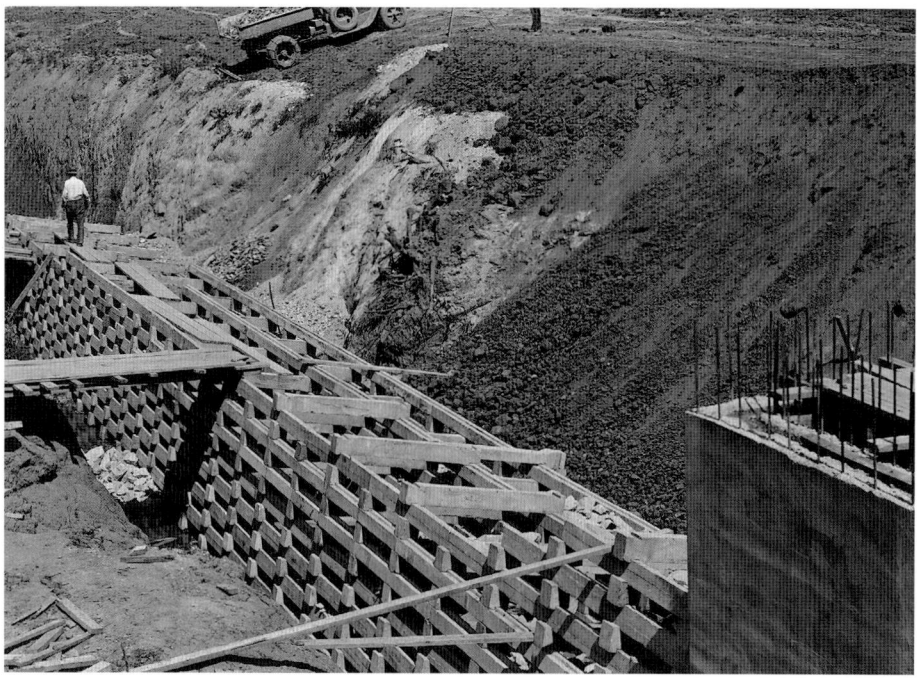

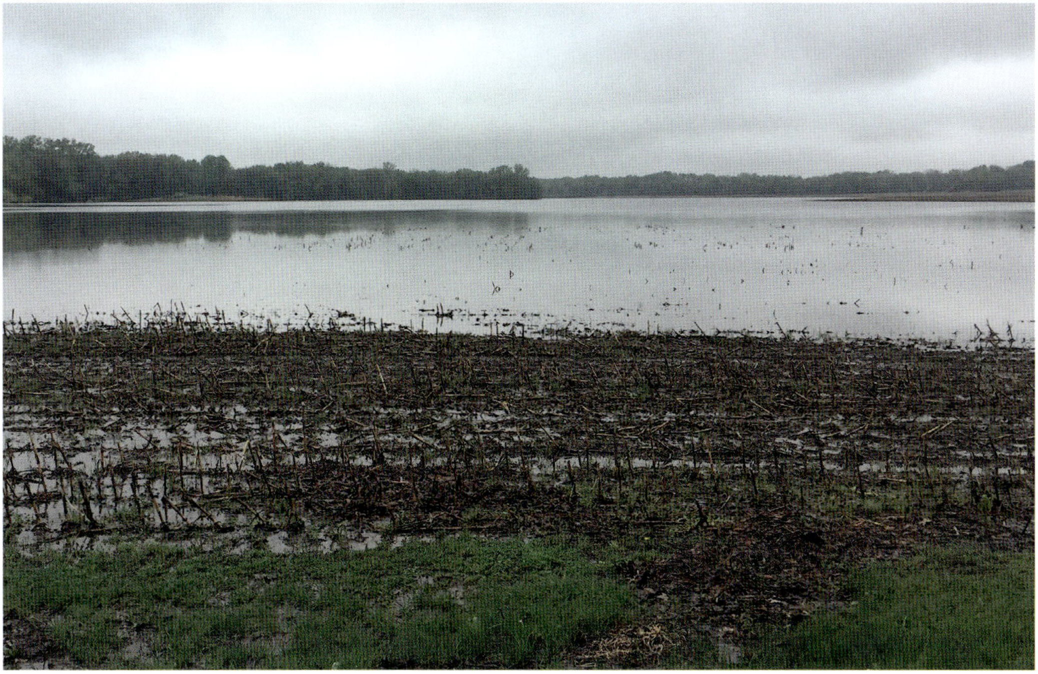

These photographs, taken during a spring flood in 2019 in the Sugar Creek Valley immediately north of the dam, offer a picture of what it was like to see the valley flood for the final time in the 1930s. Old agricultural fields gradually filled with water once the dam was completed, and old farmsteads (such as the one visible on the hill in the photo below) suddenly became shoreline property. The cornfields would never be seen again. [*Courtesy of Robert Mazrim*]

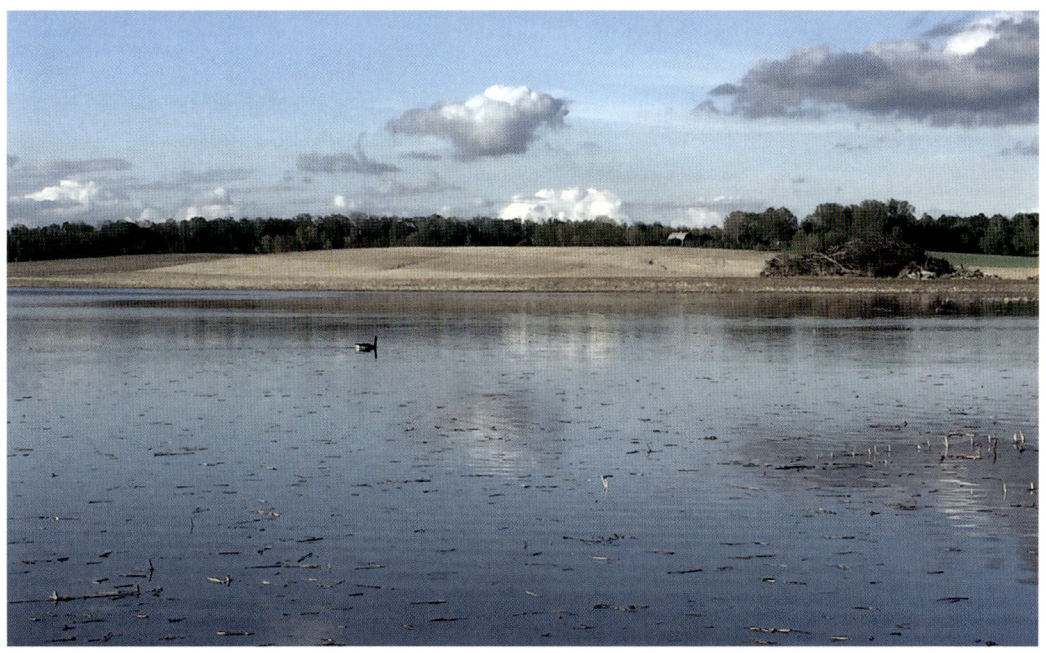

Lake Springfield in Illinois

Early in the construction of the lake, rains came early and the creek flooded too soon. Here, the flooded Sugar Creek surrounds the unfinished dam. When the dam was completed, the level of the new lake could be controlled through the raising and lowering of five separate gates (below). [*Sangamon Valley Collection, Lincoln Library*]

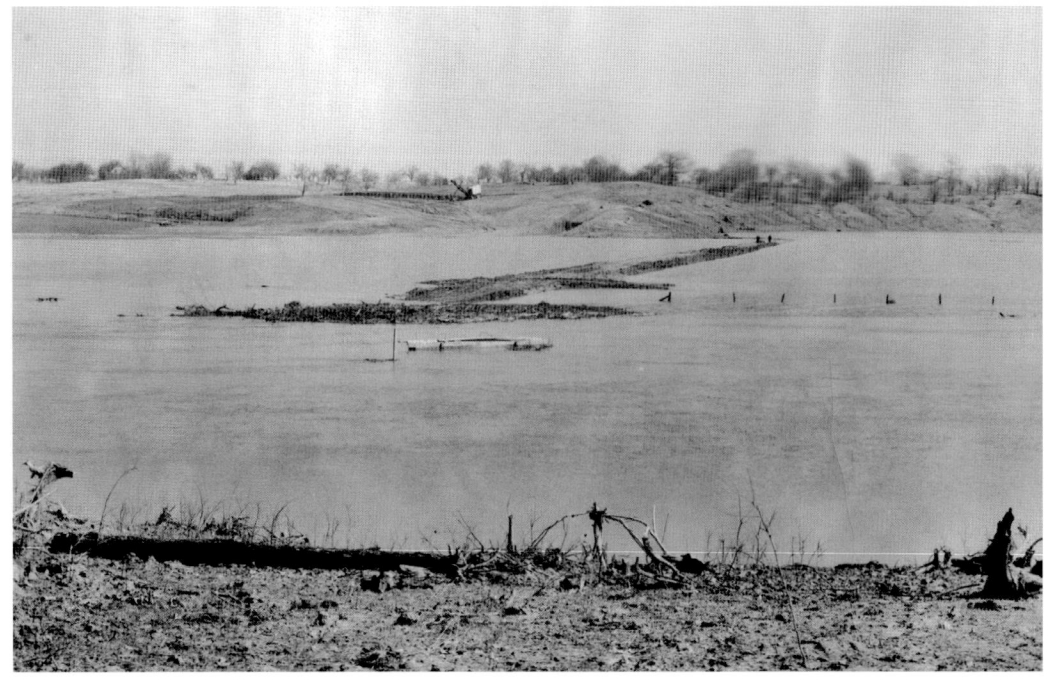

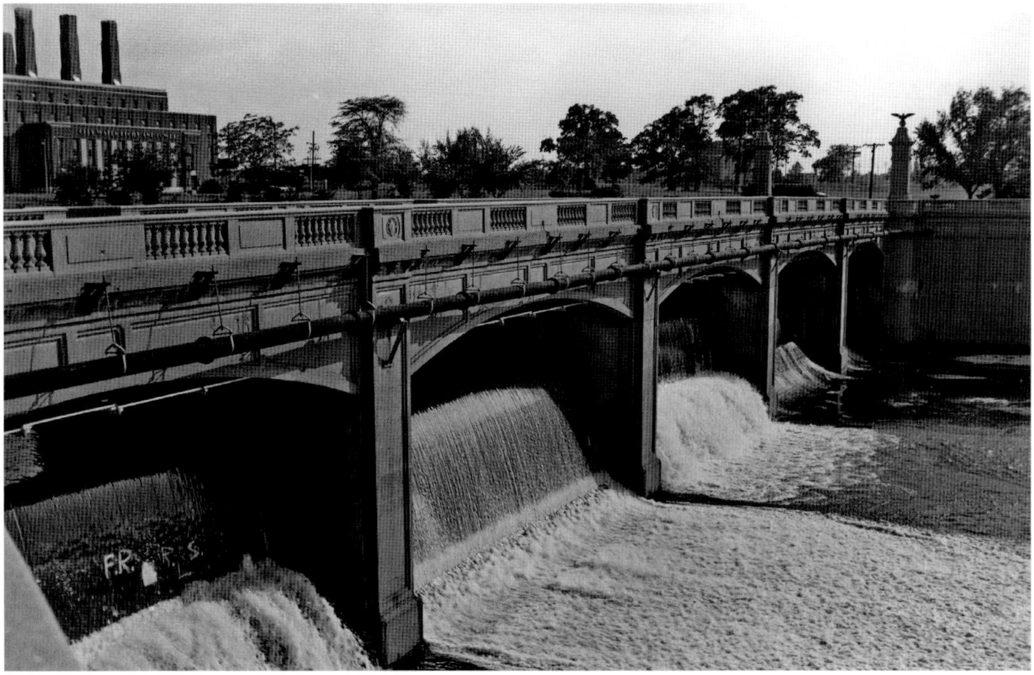

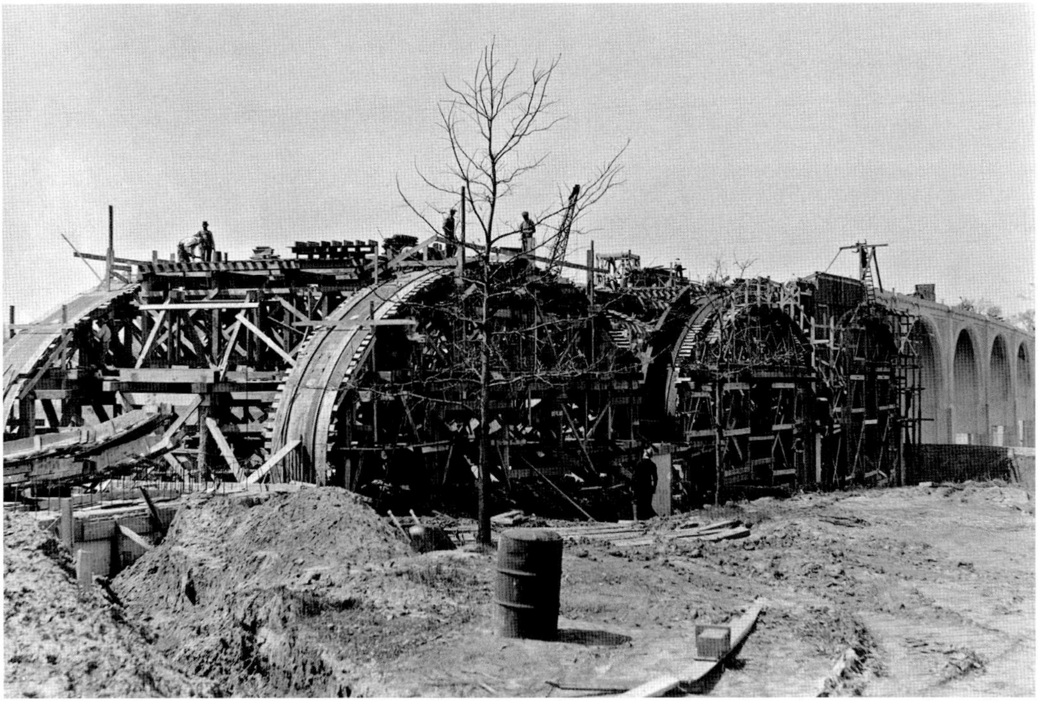

Building a Lake

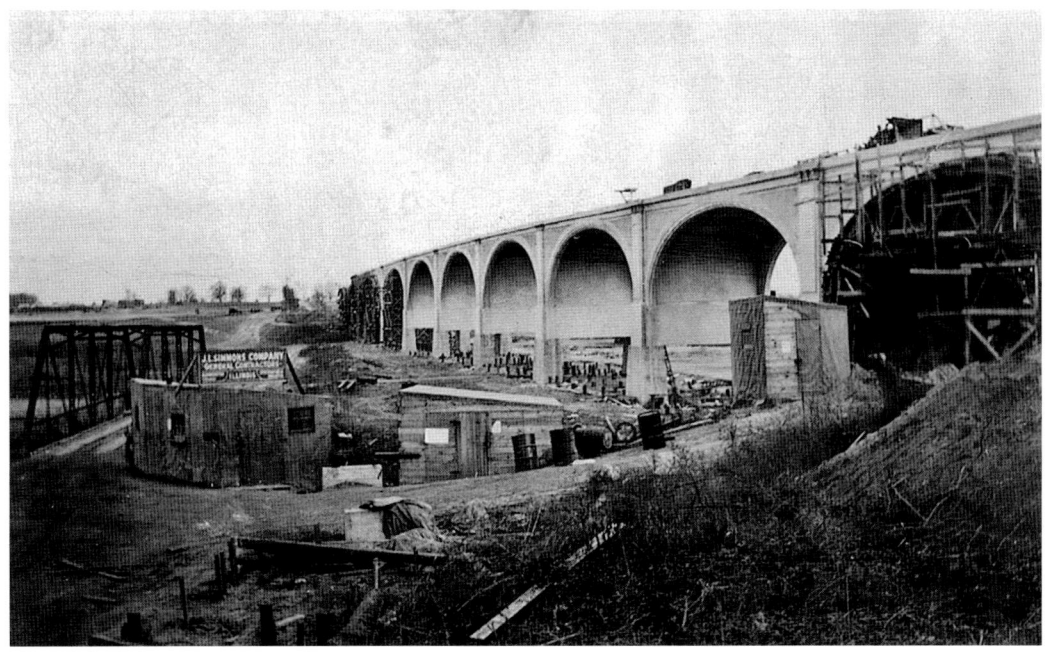

The construction of what would be called Lindsay Bridge, formally Fox Road Bridge, in the early 1930s. Note the complex forms required for casting the cantilevered concrete bridge. During the construction of the bridge, the original iron Fox Road Bridge was left in place. It was dismantled shortly afterwards. [*Sangamon Valley Collection, Lincoln Library*]

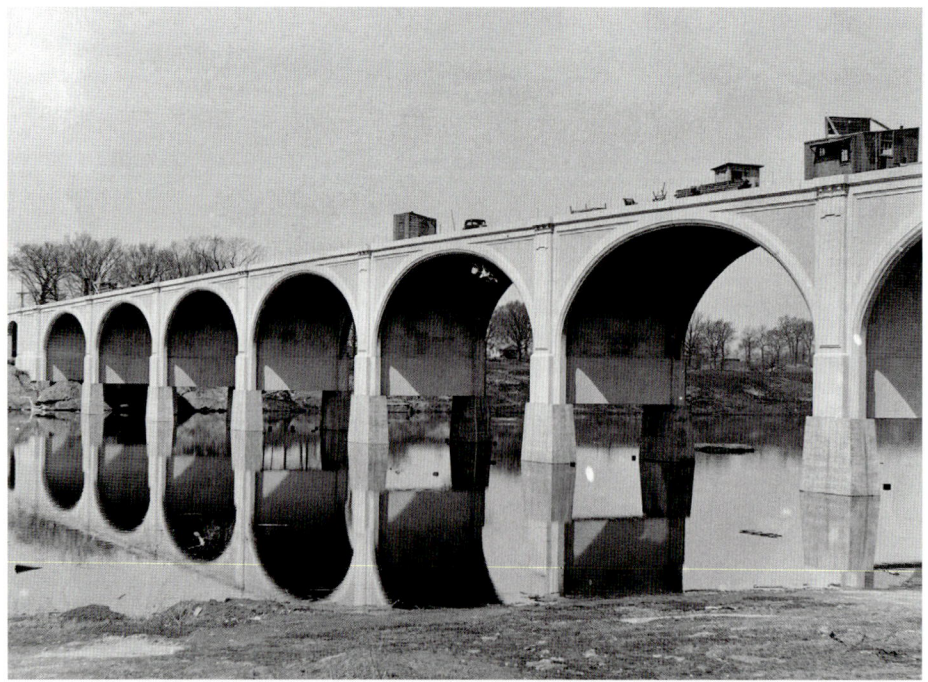

A view of Lindsay Bridge shortly after its completion, as the creek below is beginning to flood. [*Sangamon Valley Collection, Lincoln Library*] Below, the bridge as it appears today. [*Courtesy of Robert Mazrim*]

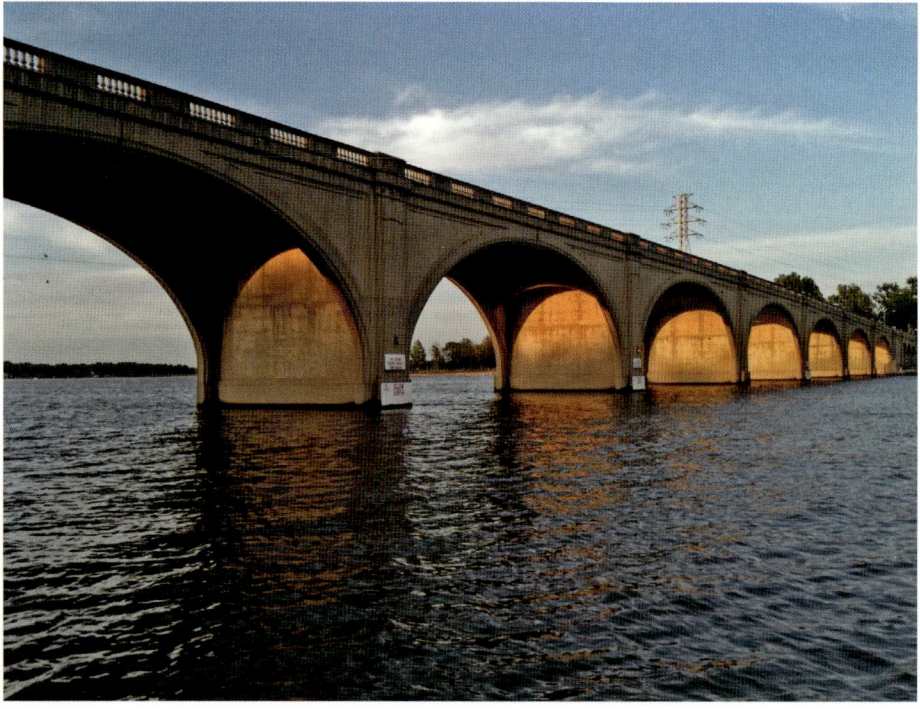

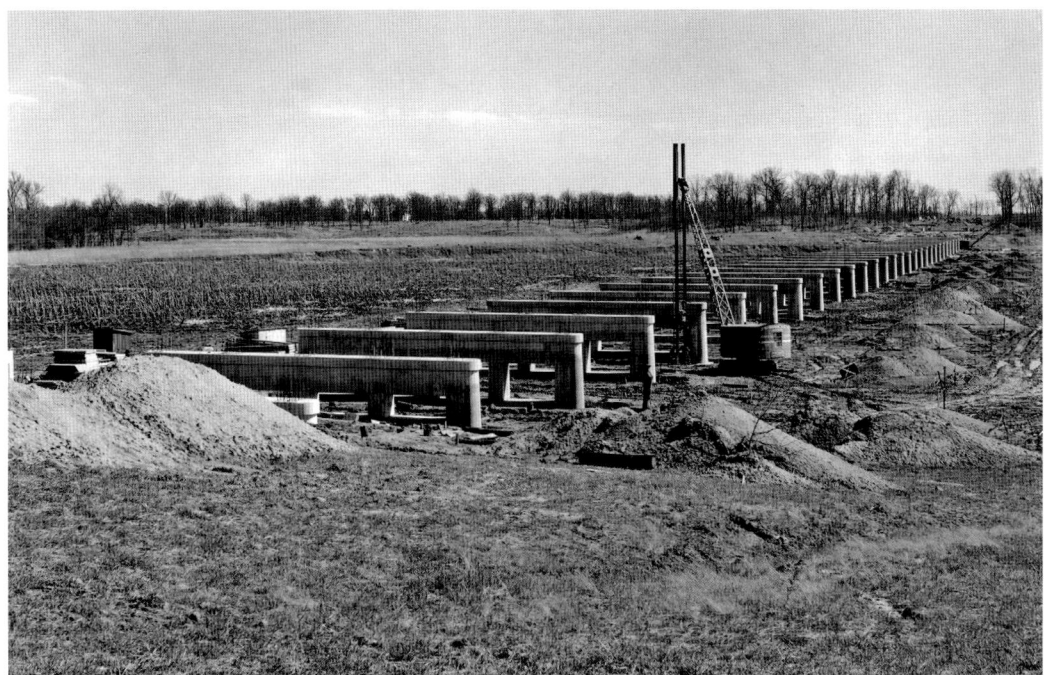

A new right-of-way was created for the legendary Highway 66 during the construction of the lake. Above can be seen early stages of bridge construction, in the middle of a cornfield that would soon become the lake. In the late 1960s, this portion of Highway 66 became Interstate 55. Most traffic between Chicago and St. Louis crosses the lake at this point. [*Sangamon Valley Collection, Lincoln Library*]

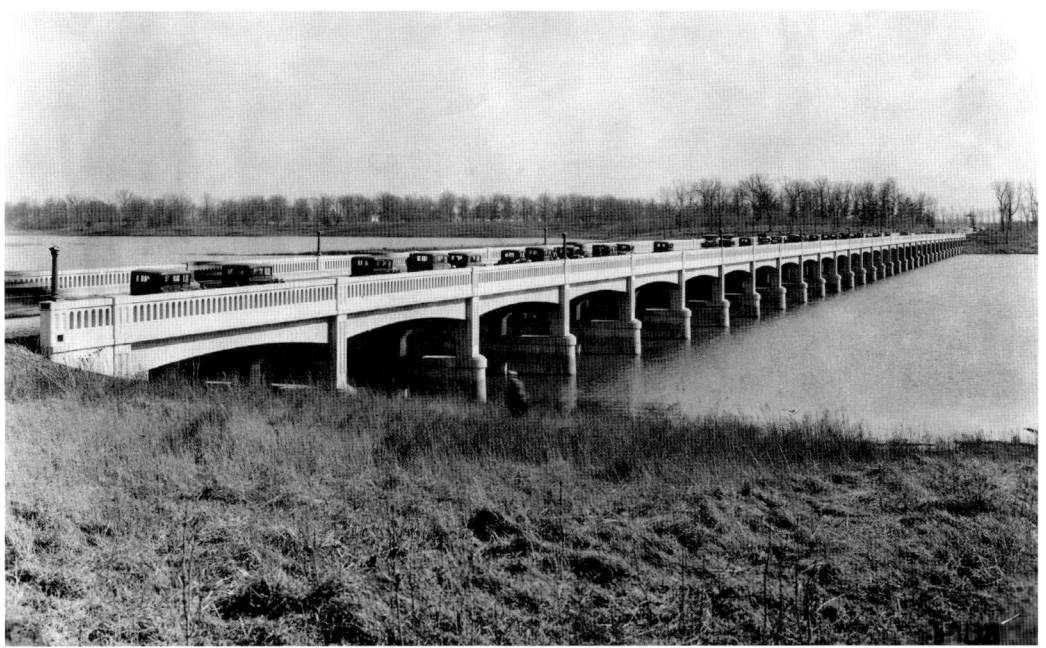

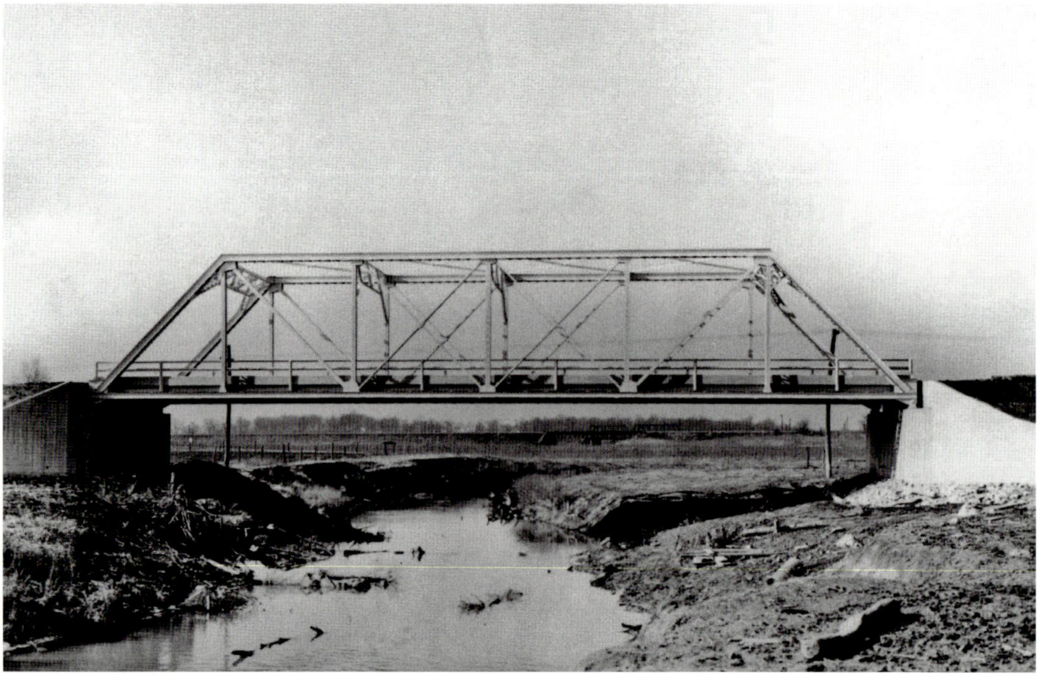

Several new bridges were required during the construction of the lake. Above is an iron bridge over the Lick Creek channel, at the extreme western end of the lake. [*Sangamon Valley Collection, Lincoln Library*] Below is a railroad bridge at the site of the village of Cotton Hill. [*Courtesy of Robert Mazrim*]

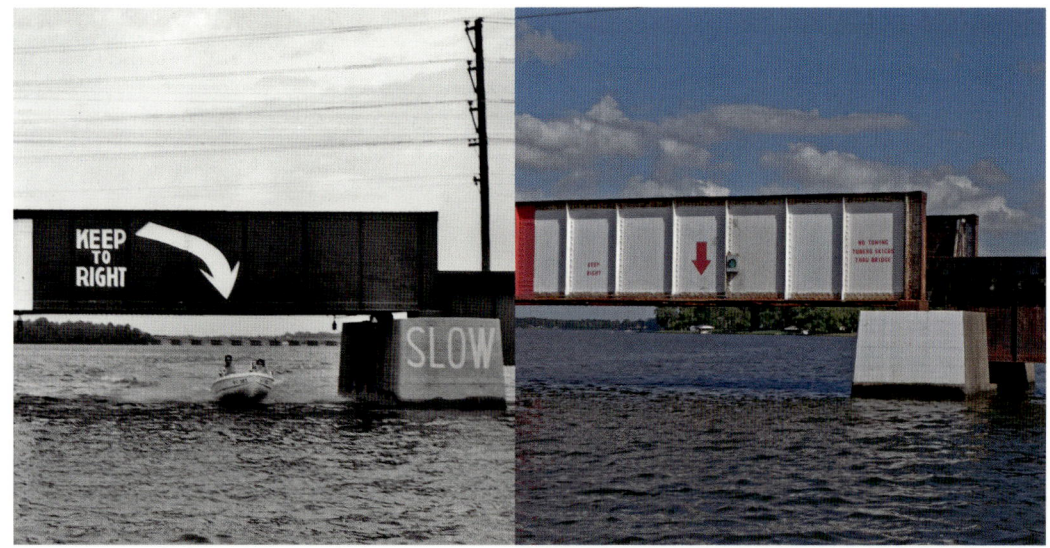

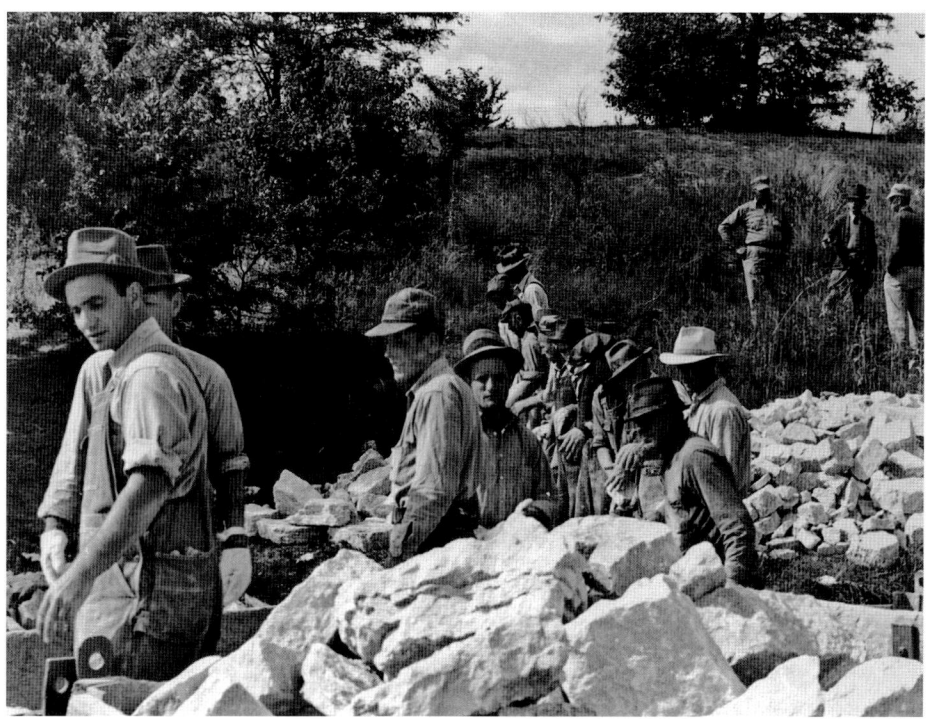

Over 57 miles of shoreline of the new lake required erosion control. Riprapping crews such as these piled tons upon tons of limestone along the shores of the lake, using barges, cranes, and manual labor. [*Sangamon Valley Collection, Lincoln Library*]

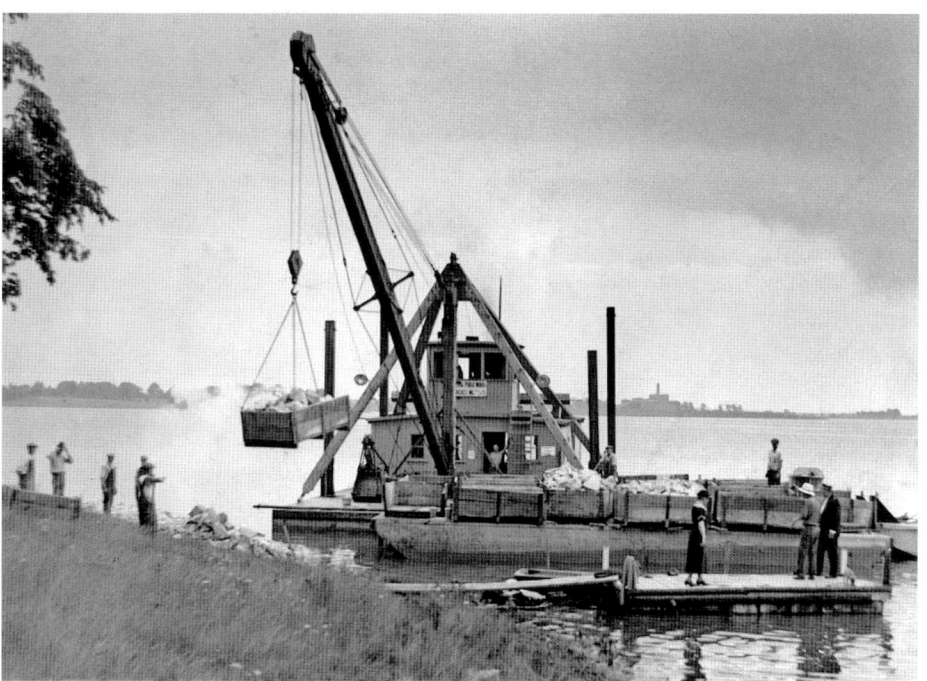

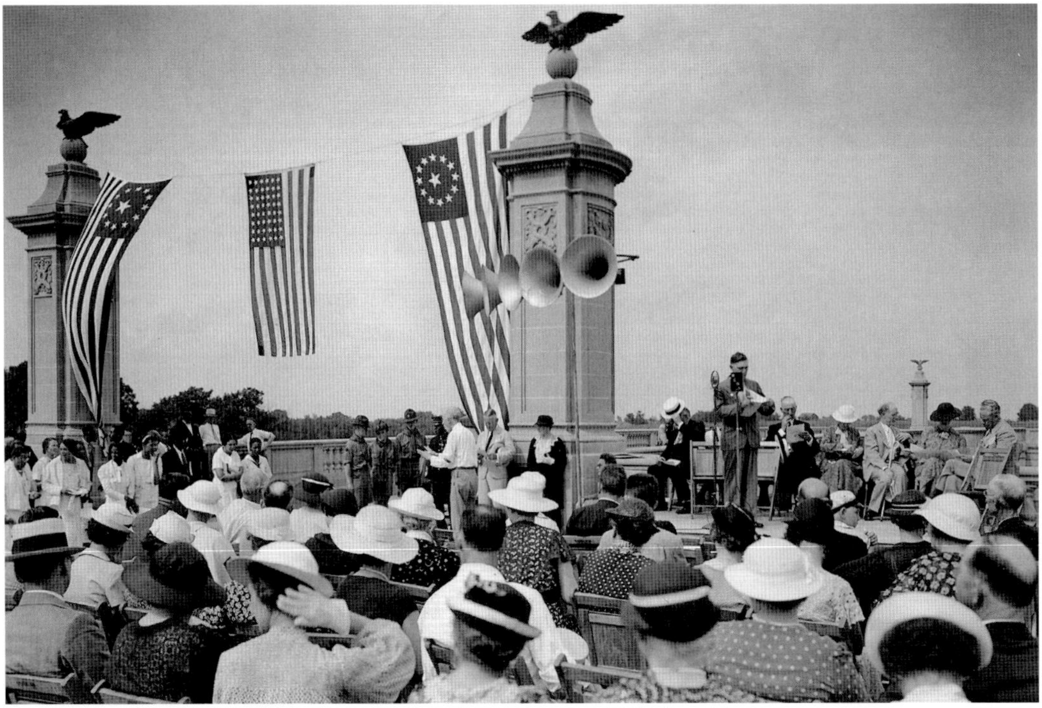

Lake Springfield in Illinois

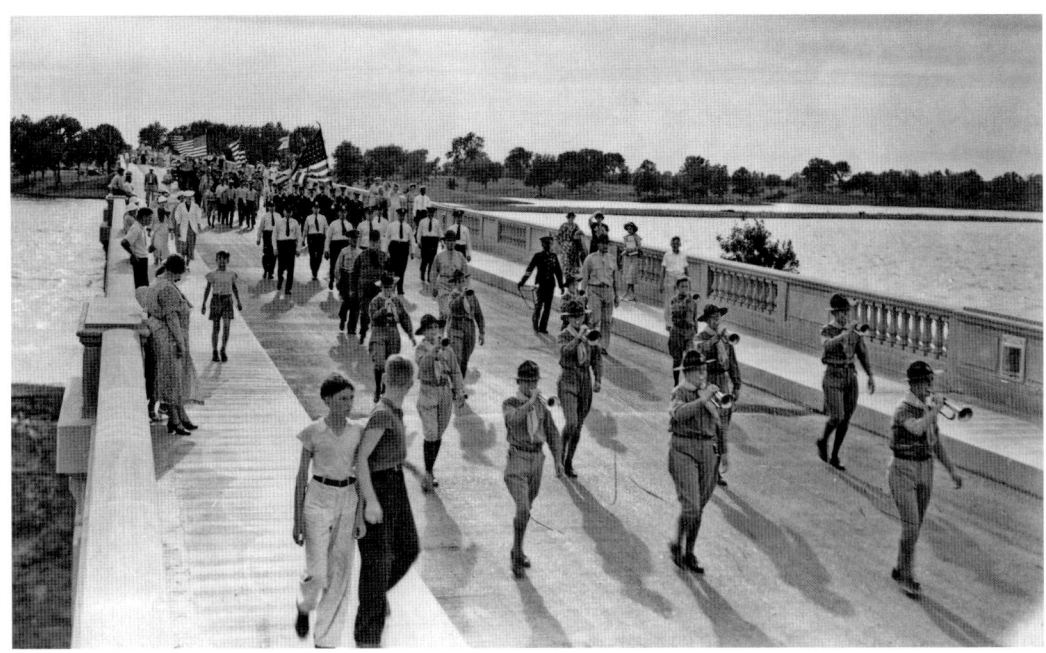

A three-day dedication ceremony for the new lake was held in July of 1935. Above is the dedication of the Spaulding Dam, and below Boy Scouts march across the new Lindsay Bridge. In attendance at the bridge dedication was poet Vachel Lindsey's widow. [*Sangamon Valley Collection, Lincoln Library*]

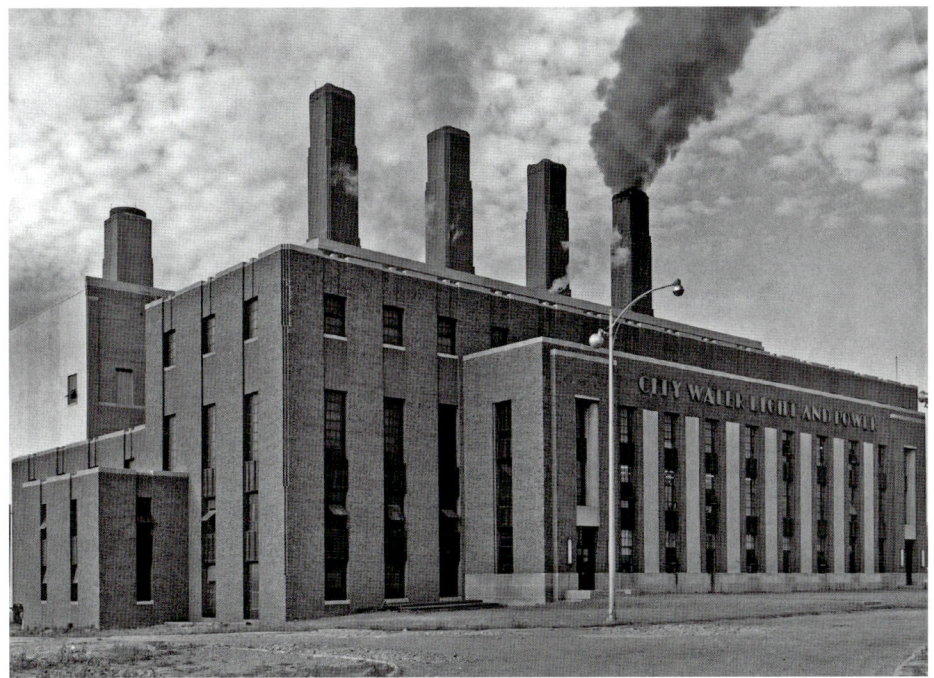

Lake Springfield was constructed not only as a municipal water supply, but also as a source of water for a new coal fired electrical plant. The new City Water Light and Power "Lakeside Plant" was dedicated in 1936. [*Sangamon Valley Collection, Lincoln Library*] The original plant has been since been abandoned, but one of the early turbines can still be seen in the decommissioned building. [*Courtesy of Robert Mazrim*]

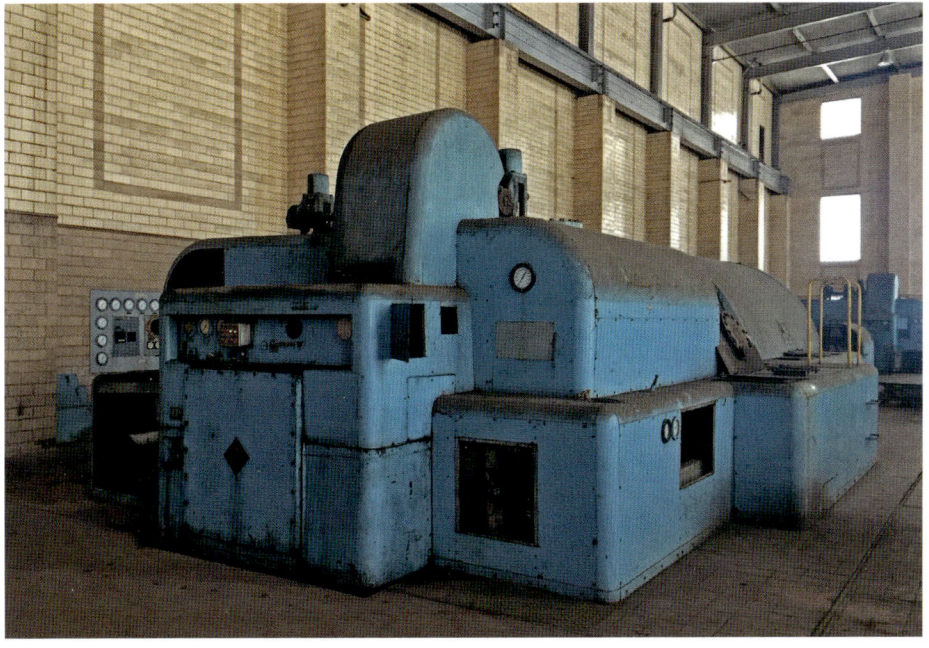

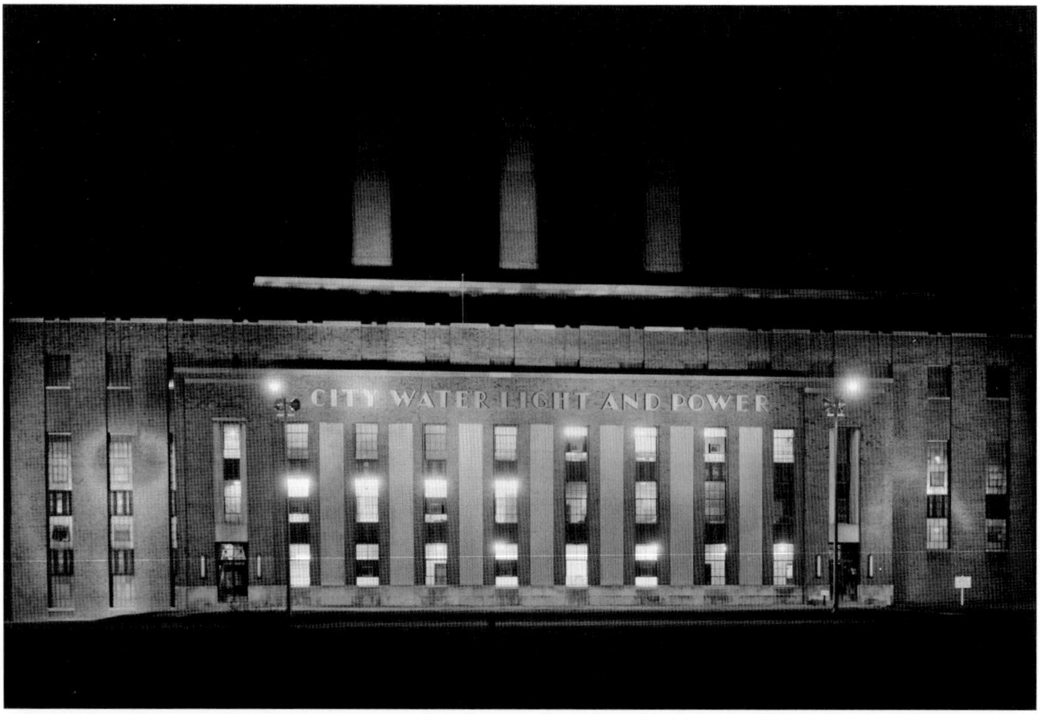

Lake Springfield in Illinois

Two views of the new City Water Light and Power plant at night, taken during the 1930s. [*Sangamon Valley Collection, Lincoln Library*]

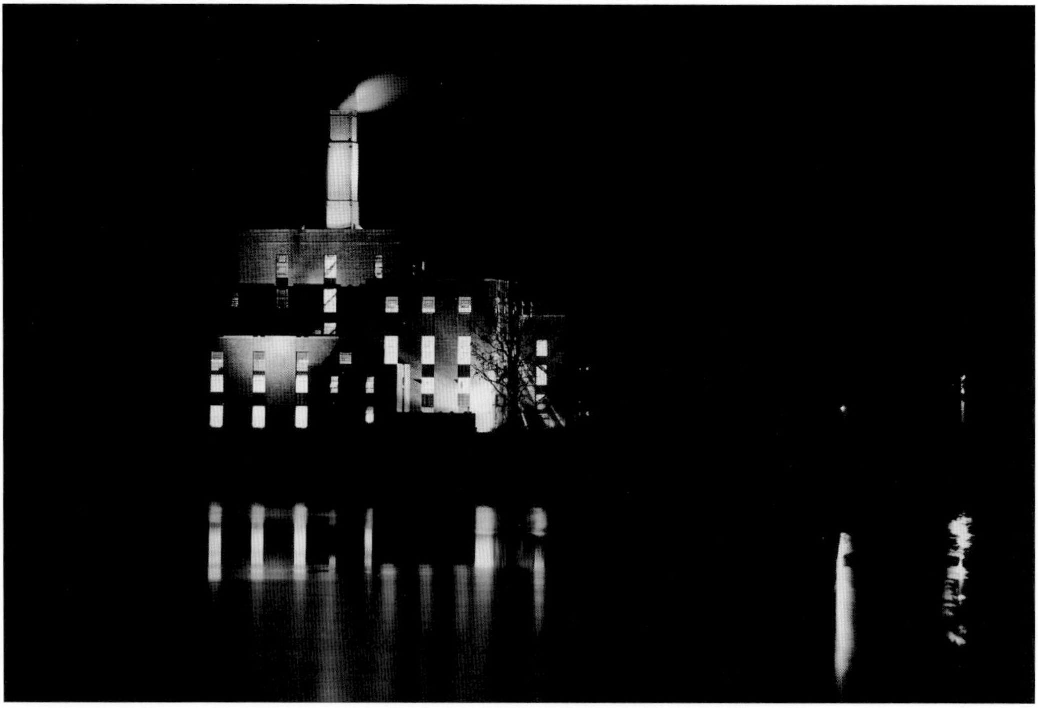

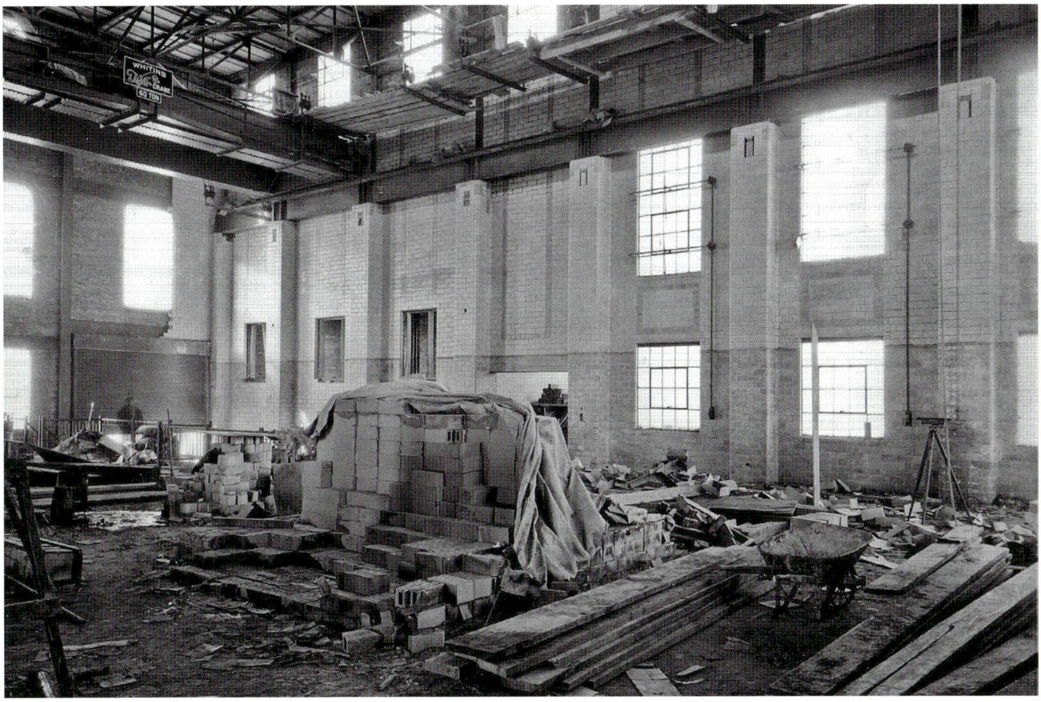

Building a Lake

A photo of the construction of the Lakeside plant during the 1930s, compared to the same portion of the facility as it appears abandoned today. [*Sangamon Valley Collection, Lincoln Library*]

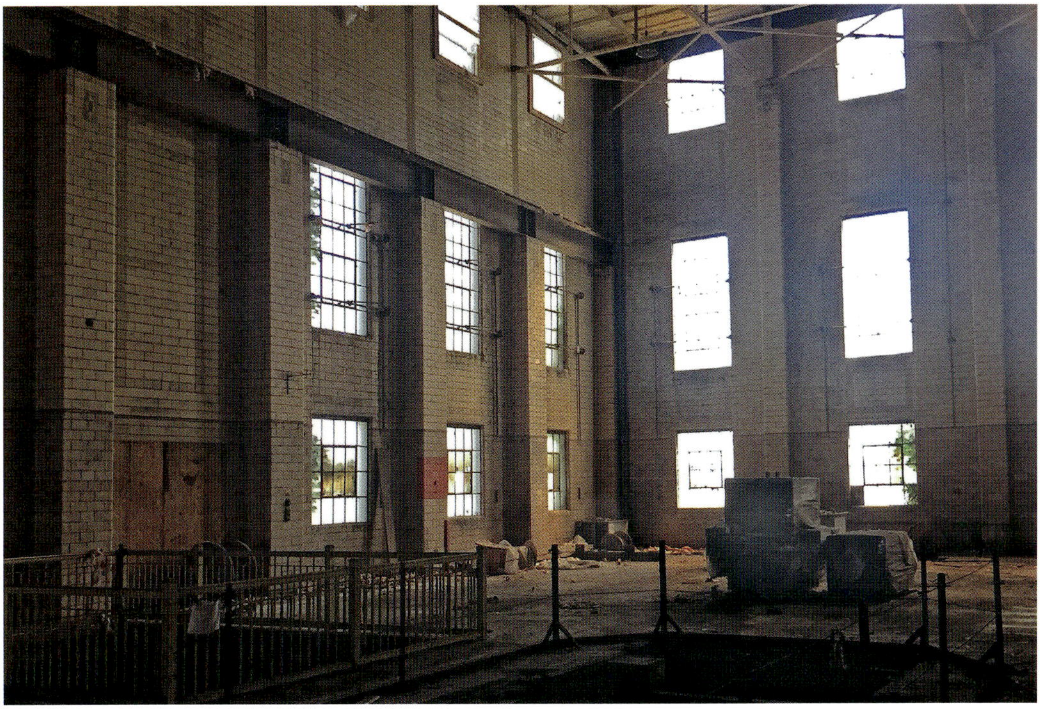

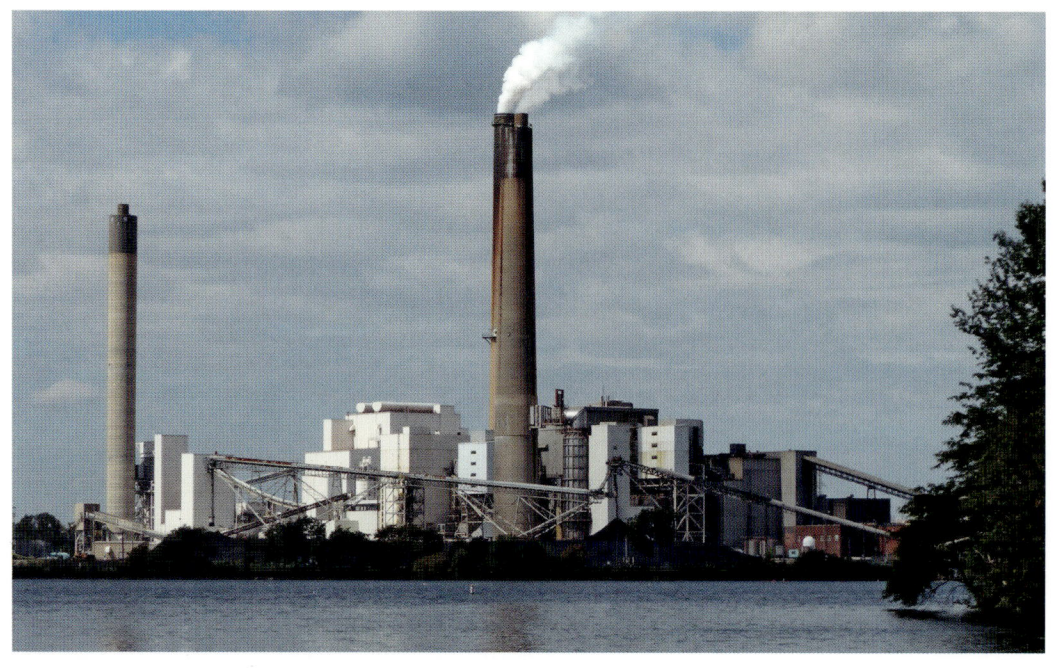

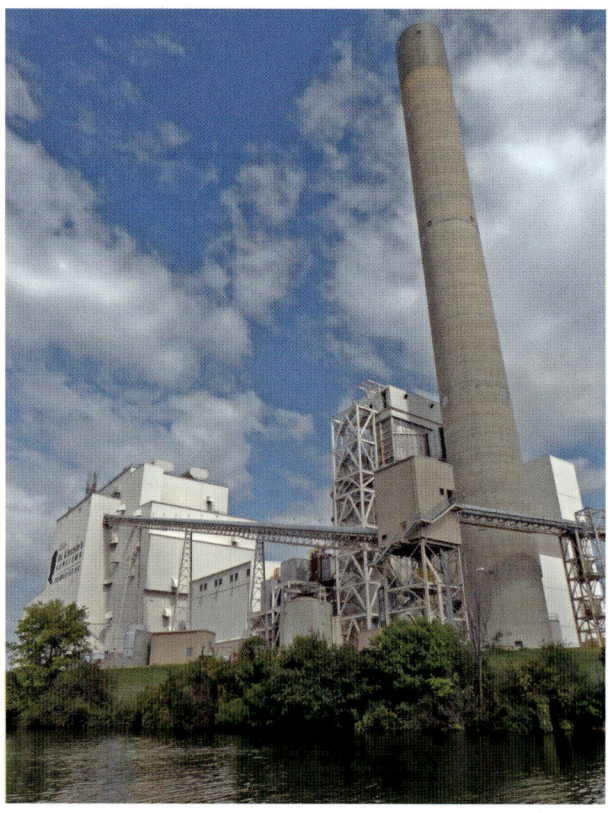

The original Lakeside plant was supplemented by several new power plants during the 1960s and 1970s (above). In 2009, the 200-megawatt Dallman #4 station was constructed (left). [*Courtesy of Robert Mazrim*]

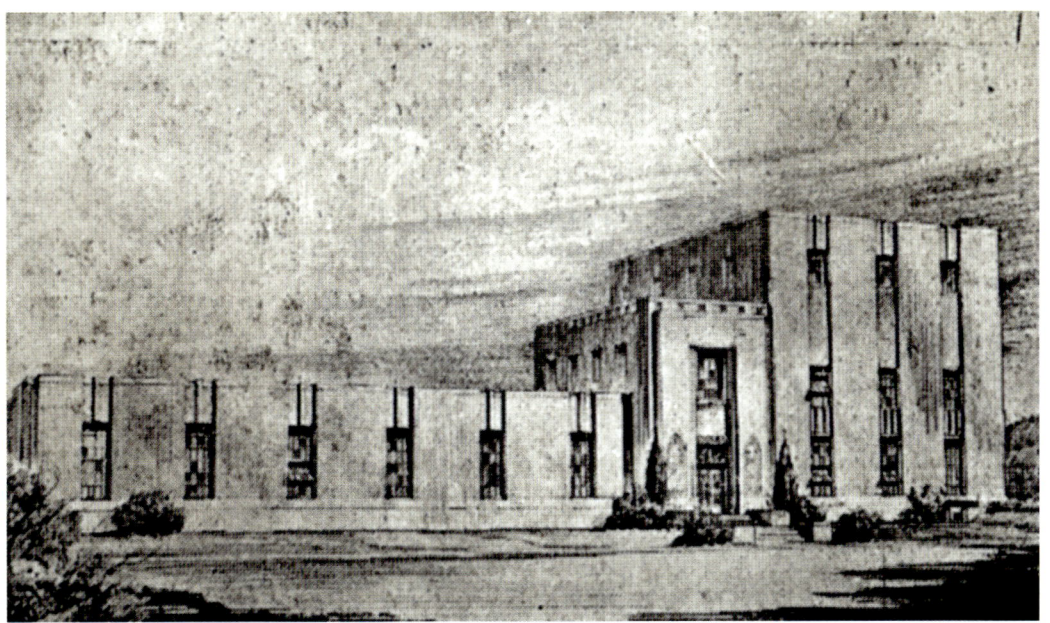

Original architect's rendering of the municipal water treatment plant, which was completed in 1934, shown with a view of the building as it appears today. [*Above: Sangamon Valley Collection, Lincoln Library; Below: Courtesy of Robert Mazrim*]

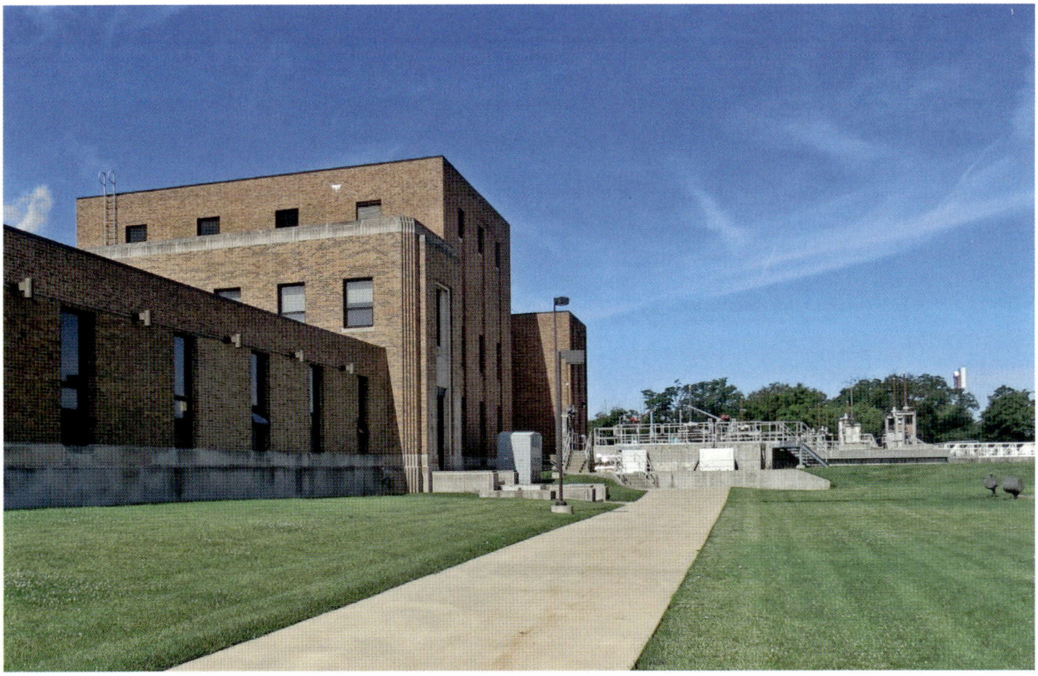

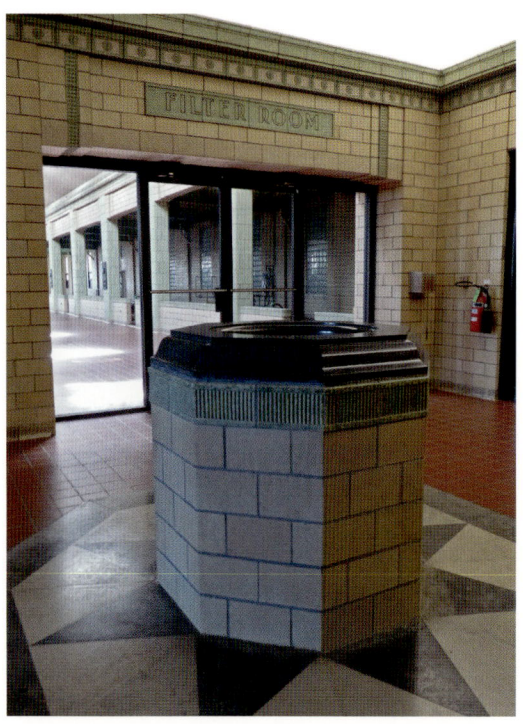

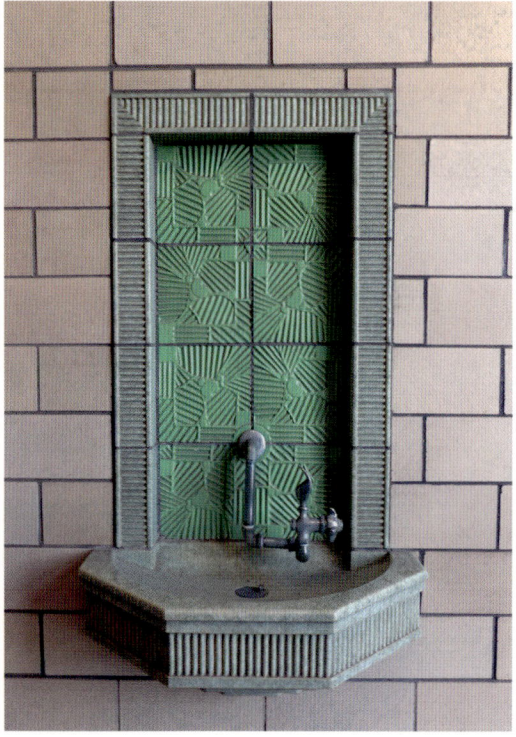

Inside the water treatment plant is the well-preserved decorative tile work from early 1930s. Immediately inside the main public entrance of the building is an impressive art-deco water fountain, which encouraged visitors to "sample the product" of the new treatment facility. [*Courtesy of Robert Mazrim*]

Right: Charles Spaulding, designed and patented the water clarifier used at the new treatment facility at Lake Springfield. [*Sangamon Valley Collection, Lincoln Library*]

Below: In 1975, artists Saunders Schultz and William Severson created a sculptural water fountain on the campus of Sangamon State University (now University of Illinois at Springfield), which was inspired by Spaulding's design and patent. [*Courtesy of Robert Mazrim*]

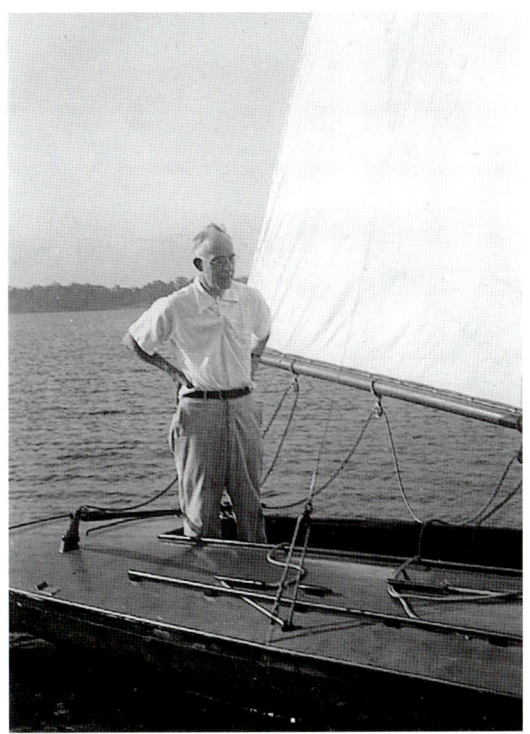

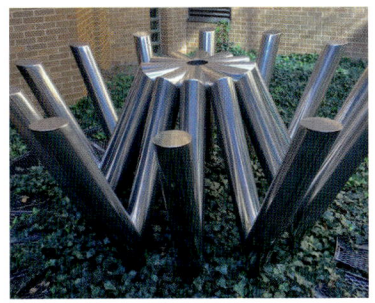

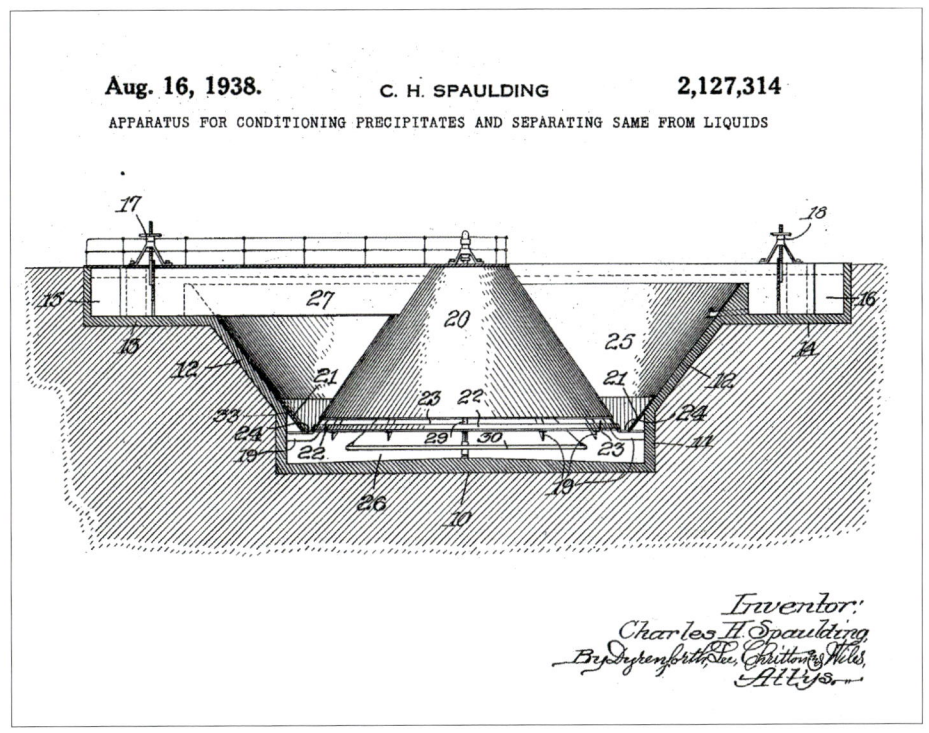

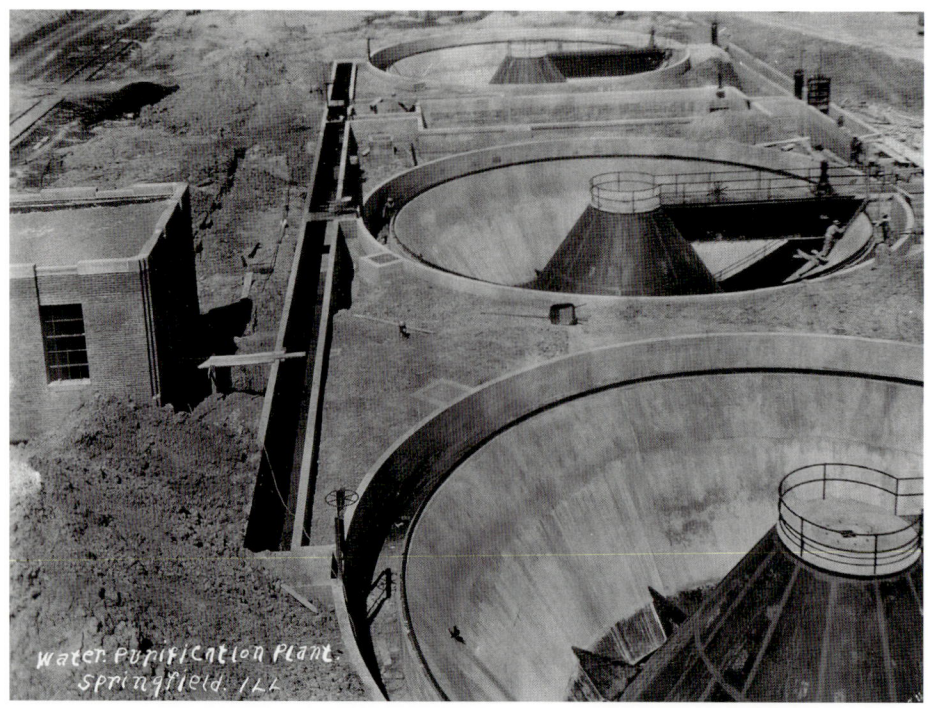

The water purifiers as they appeared in 1936, and how they appear today. [*Above: Sangamon Valley Collection, Lincoln Library; Below: Courtesy of Robert Mazrim*]

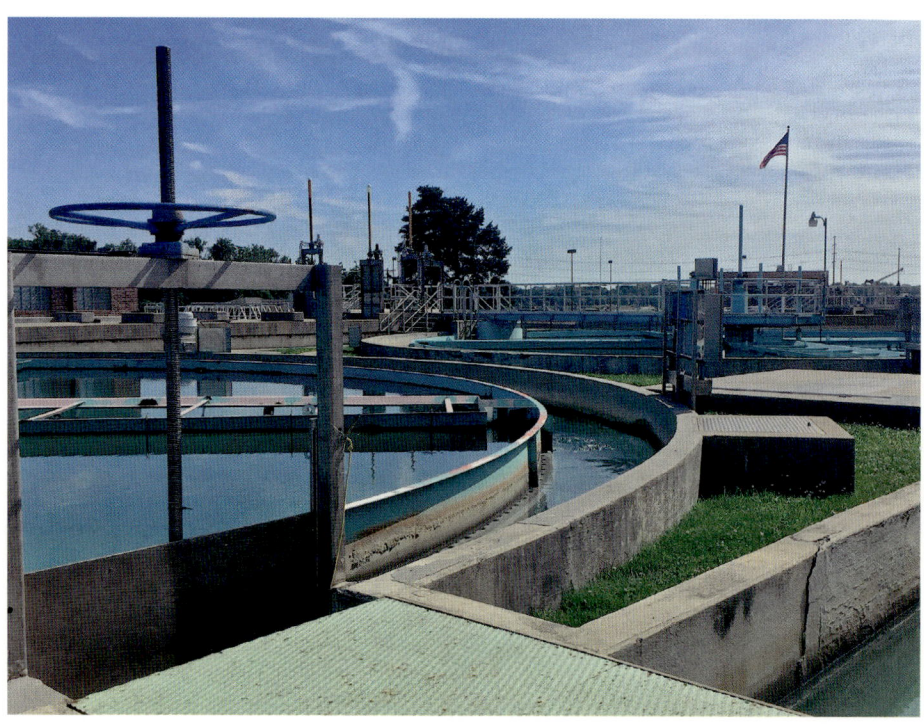

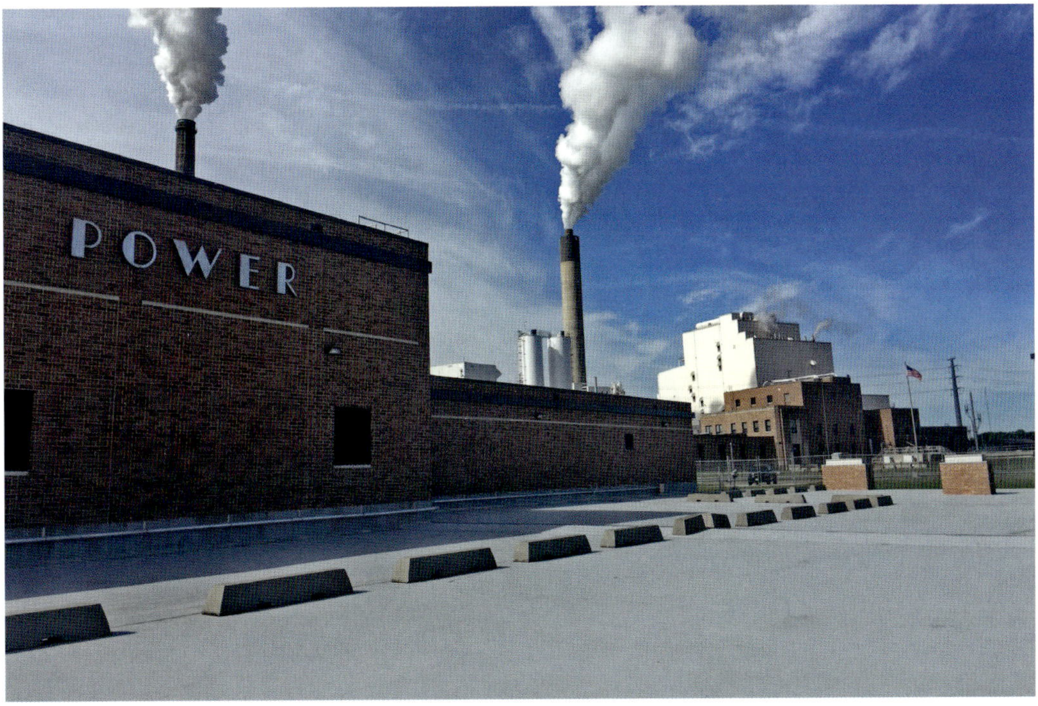

Building a Lake

A large concrete slab at the front of the CWLP property is actually the "lid" to a massive underground storage facility for water, to serve as a back-up supply in times of drought or emergency. [*Above: Courtesy of Robert Mazrim; Below: Sangamon Valley Collection, Lincoln Library*]

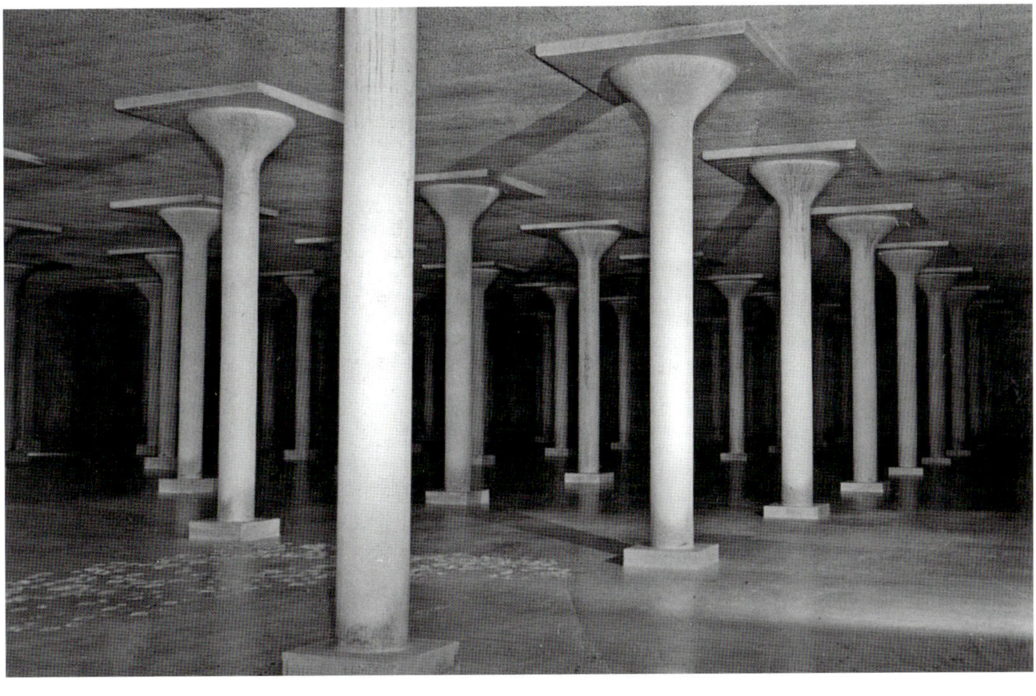

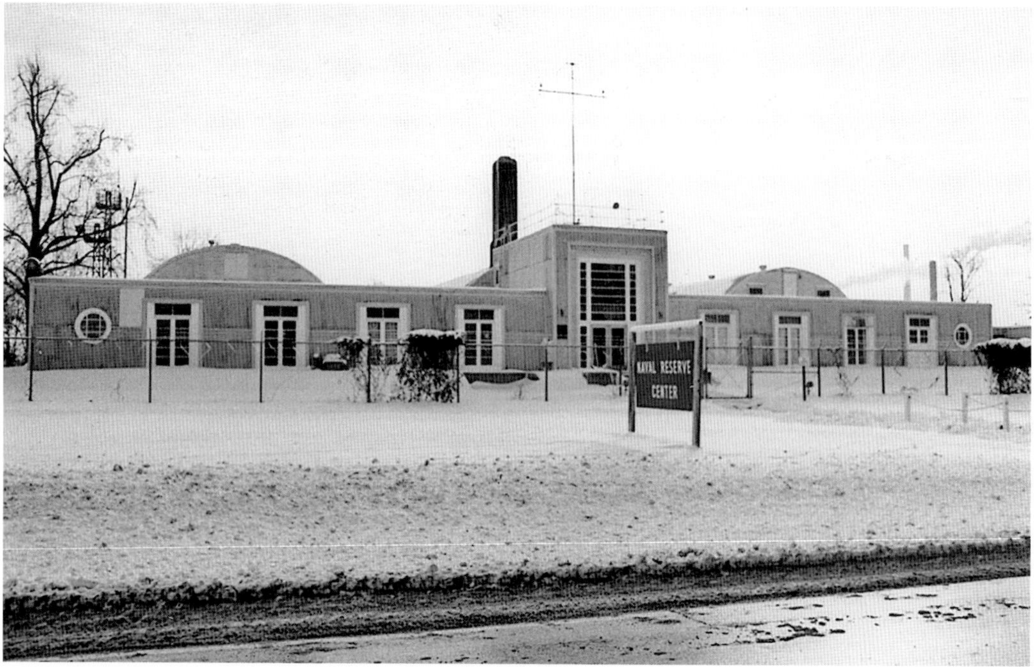

In 1942, the Illinois National Guard constructed a barracks near the Spaulding dam to provide protection to the power and water supply during World War II. After the war, the building was remodeled to serve as a Naval Reserve training center. [*Sangamon Valley Collection, Lincoln Library*]

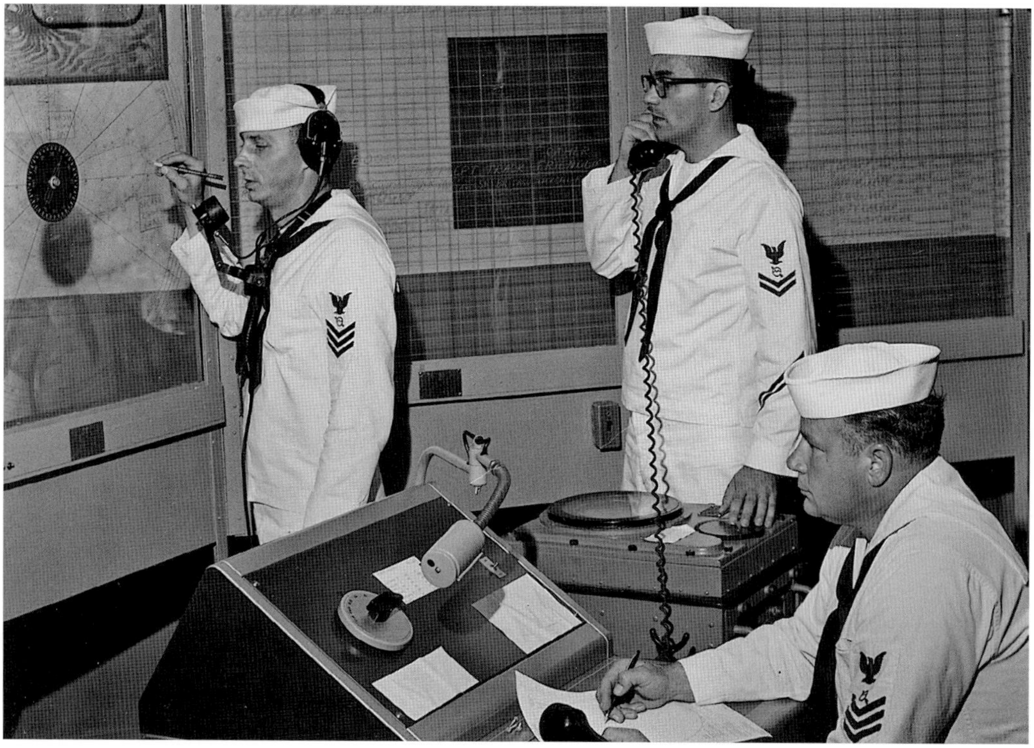

4

New Neighborhoods on the Water

The land surrounding the new reservoir was also purchased by the city, and was divided into over 735 residential lots, as well as spaces for parks, private clubs, and other municipal facilities. Two new principal roads encircled the lake—East and West Lake Shore Drive. Part of East Lake Shore Drive, on the southern edge of the lake, actually follows the same course as the ancient Edwards' Trace trail. From these roads were smaller "lanes" that provided access to the subdivided residential lots along the shoreline. Between those lanes and the main roads were green spaces that provided for a more natural surrounding and for wildlife habitat. Hunting was banned on all natural areas of lake property. Most of the planned green spaces are still intact today and have not been developed. Some even feature restored prairie grass.

Lots along the shoreline of Lake Springfield were not sold outright, but instead, homebuilders were issued sixty-year leases ranging from $60 to $300 per year. Lake leases were used to distribute the properties for two reasons: due to a state law that forbid the city from selling city property without passing ordinances, and also to maintain a certain amount of oversight and protection of land that was adjacent to the water. Covenants required that new homebuilders spend pre-established minimums on the construction of new homes—ranging from $1,000 (or about $50,000 in today's relative value) to $6,000 in certain areas of the lake. The sixty-year lease essentially assured buyers that the property would be secure for the reminder of their lifetimes. Later, the lease period was extended to ninety-nine years.

The city provided 175,000 seedling hardwoods and fruit trees from a nursery located on the north shore (along what is now Stevenson Drive), in part to encourage erosion control. Riprapping services were also offered at no cost to homeowners along the water, to stabilize the shoreline. At first, this consisted of quarried limestone that was hauled to the shorelines on barges and put into place by WPA laborers or teenagers working in summer jobs. This service was initially free to landowners, as it provided an important stabilization for the reservoir itself. Later, the shorelines were sometimes repaired with fragments of concrete slabs salvaged from construction projects, including fragments from portions of Route 66 that removed during the construction of Interstate Highway 55 during the 1960s. Finally, during the 1980s, landowners begin replacing the rubble stone shorelines with more permanent steel seawalls.

A number of houses along the new shoreline actually predated the lake—having been farmhouses that once overlooked the floodplain. Approximately two dozen such homes were sold and remodeled as lakefront homes during the 1930s, and several still stand today.

The first new houses constructed on the lake were built near what was still known as Fox Bridge Road, which crossed the lake over the new Vachel Lindsey Memorial Bridge. The first house was described as a "modern cottage," on a point south of the bridge.

The second house, built in 1933, had a unique beginning. The house was constructed by the Knights of Columbus, and christened "Lakeside Villa." It was donated by the K of C as a raffle prize during its annual barbecue. The winner of the house, a twenty-one-year-old candy store clerk from Danville, Illinois, accepted cash instead of the house. Lakeside Villa was later purchased by Springfield Mayor John W. Kapp as a Christmas present to his wife.

By the 1980s, the neighborhoods surrounding Lake Springfield would become known for their large houses and affluent families, but the initial community that grew up around the lake was a generally a middle-class one. Many of the homes were quite modest and even included small weekend cabins. In Cottage Grove Lane, several houses were actually constructed from converted Quonset huts, used to house workers while they were building the lake during the early 1930s. Even today, a couple of houses in this neighborhood feature the distinctive arched ceilings from the Quonset huts that were later surrounded by additions. In other lanes, small homes built of logs were meant to suggest cabins or resort houses. One of these, built in Orchard Lane, was constructed of recycled telephone poles. Because the poles were treated with creosote to prevent rot, the neighbors often remarked on the pungent scent of oil that surrounded the house on hot summer days.

Architecturally, the styles of first houses that were built along the lake before World War II reflected styles popular with middle-class families in the Midwest. These included small cottages and larger American "foursquare" or traditional "farmhouse" styles with Colonial, Classical, Federal, or Gothic detailing. These often included multiple gables and plans that featured views of the water. After the war, ranch houses became very popular, but these too were stylized and included historical or revival detailing (Cape Cod being a popular style along the water). Also introduced after the war were new Modernist or International styles.

The description of homes included in a "Life at Lake Springfield" house tour conducted in the fall of 1948 offers a hint of the intended character of lake homes built before the war. The homes selected for the tour included an "English cottage," a "Farmhouse of 1948," a "Modern Cottage," a "House to Entertain In," and a "House for a small family." By the end of 1948, 249 families resided on the shores of the lake, and forty-five new homes were under construction. Today, over 700 private residences dot the shoreline of Lake Springfield.

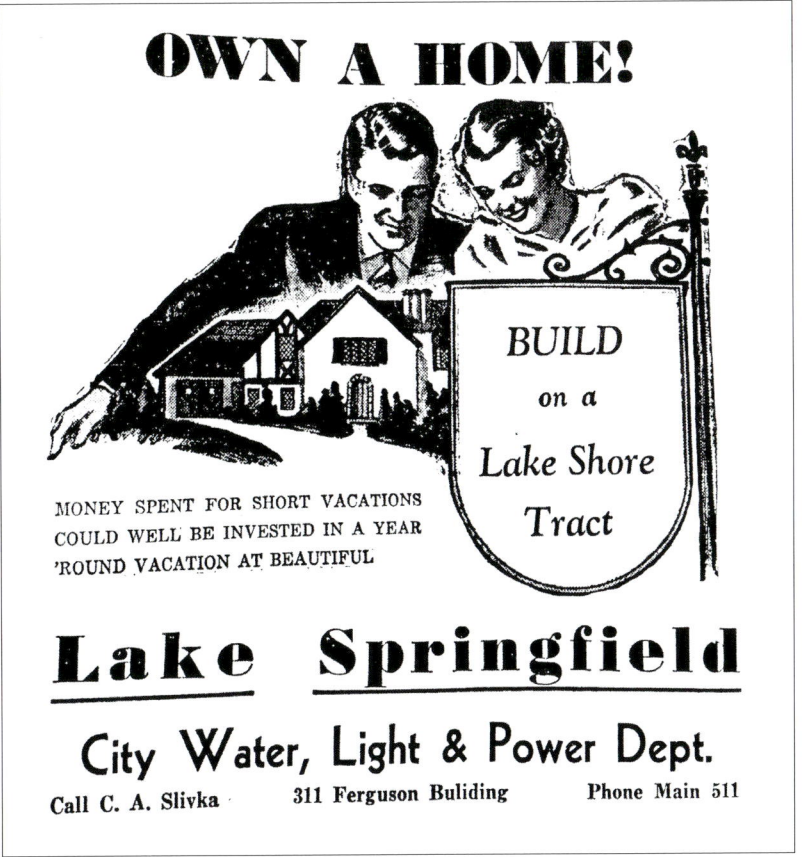

A 1930s advertisement for new home construction along the lake shore stressed that money spent for vacations could be invested in a "year-round vacation." Below is an architect's rendering and current photograph of the second house built on the Lake Springfield: "Lakeside Villa" constructed and raffled by the Knights of Columbus in 1933. [*Sangamon Valley Collection, Lincoln Library*]

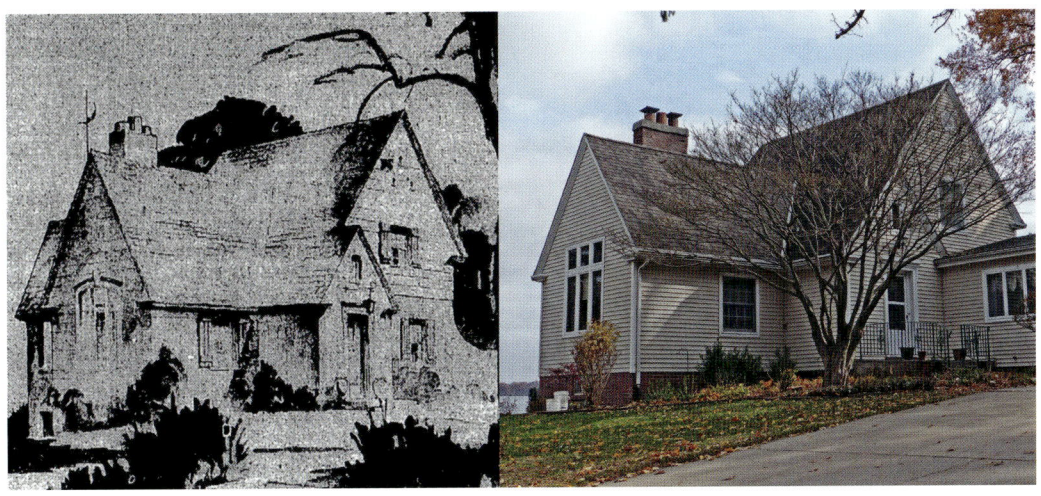

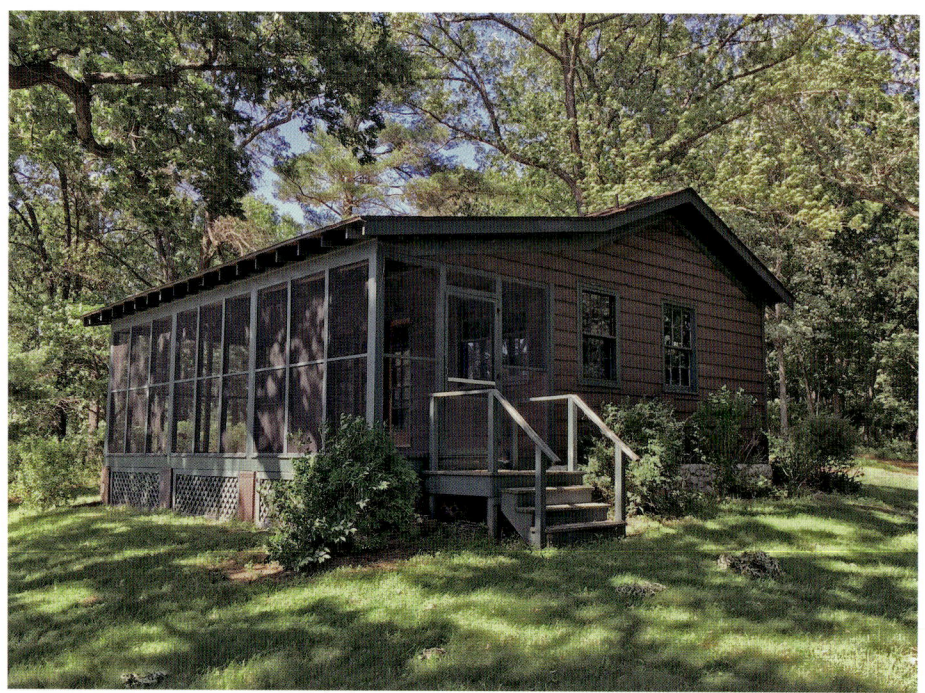

A range of styles and sizes of homes was constructed on the new shoreline. Some were designed as weekend cottages or cabins. One of those early cabins still stands today in good preservation on the far western shore of the lake. Below is the rather ramshackle construction of the city-owned cabin built by Commissioner Willis J. Spaulding for his own use. Subsequent city officials had personal use of the property until it was demolished in 1975. [*Above: Courtesy of Robert Mazrim; Below: Sangamon Valley Collection, Lincoln Library*]

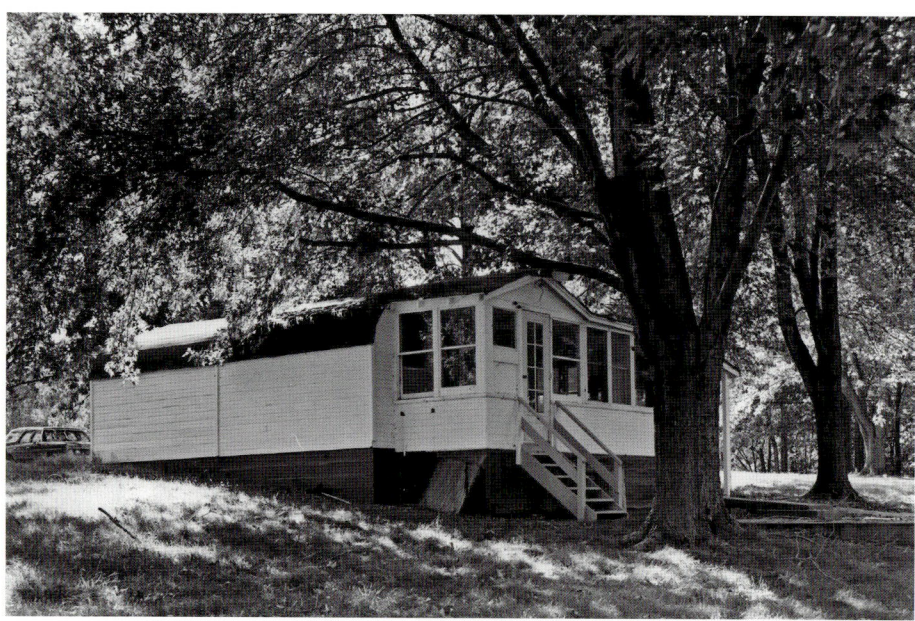

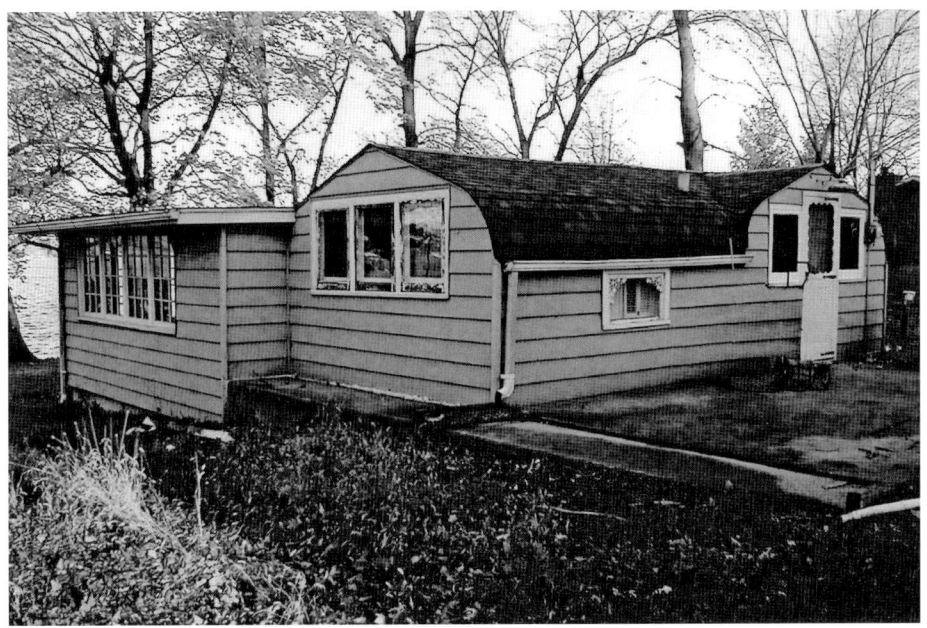

In "Cottage Grove Lane" were a series of rather small homes or cottages, including several that were built from reused Quonset huts that had been occupied by workers during the construction of the lake. Many of these were intended as weekend getaways, but were gradually enlarged for permanent residences. This example shows the characteristic Quonset hut profile from both exterior and interior views. [*Courtesy of Robert Mazrim*]

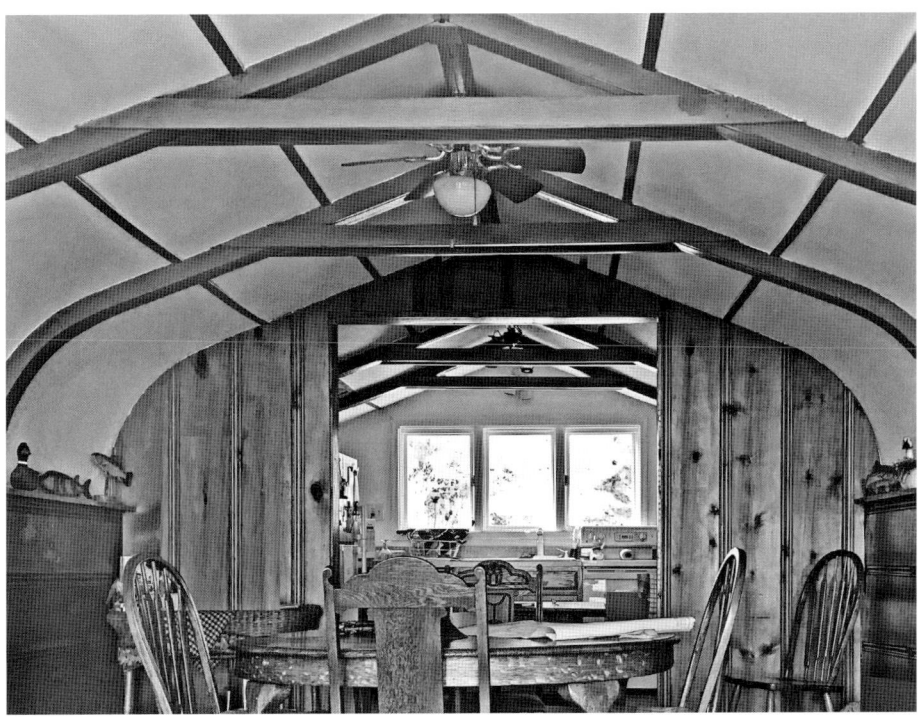

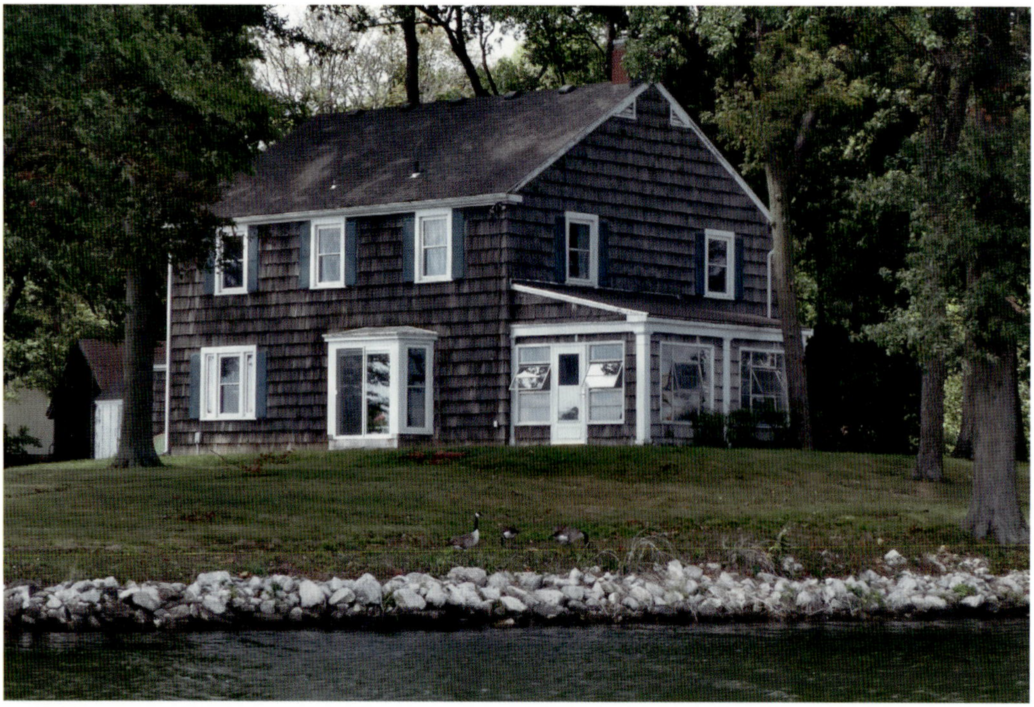

A variety of early architectural styles on the lake also included larger homes constructed for year-round residency. These Cape Cod-style homes represent an architectural reference that was popular in the Midwest during the early twentieth century. [*Courtesy of Robert Mazrim*]

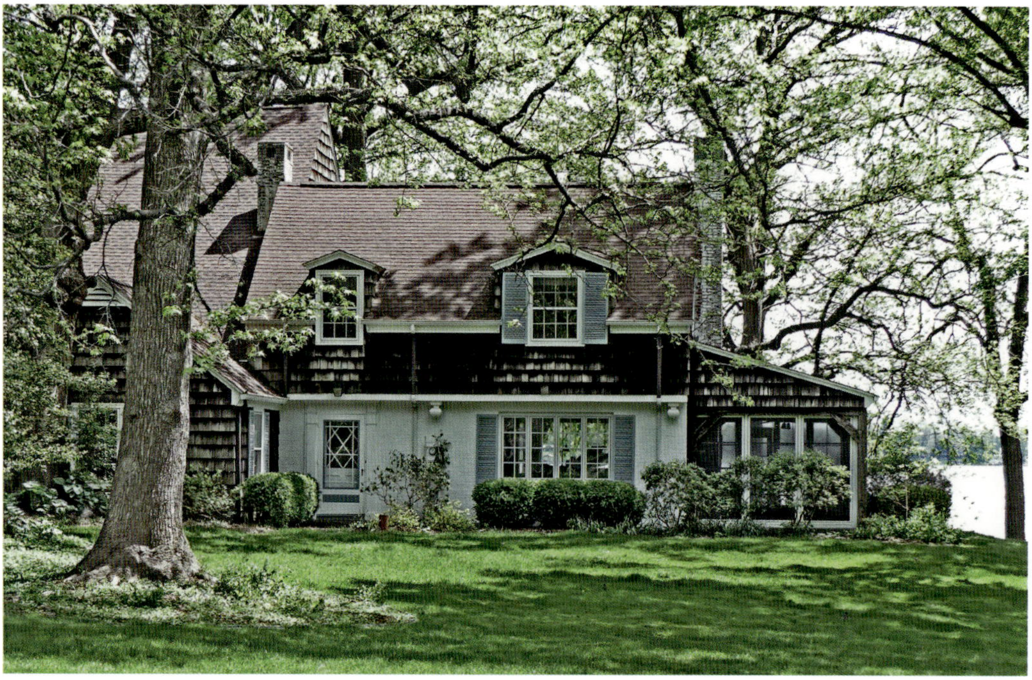

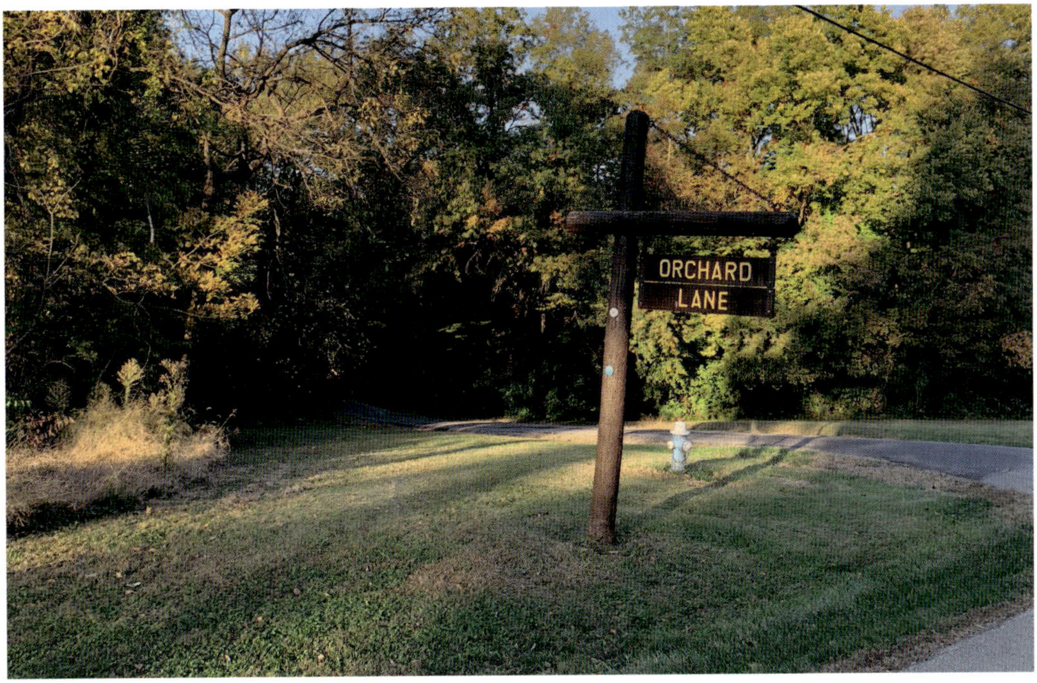

Most of the houses built along Lake Springfield are located on small loops or lanes that encircle the lake to access the waterfront properties. Those lanes include not only residential lots, but also natural, undeveloped spaces as well. Orchard Lane includes a large area of native prairie restoration. [*Courtesy of Robert Mazrim*]

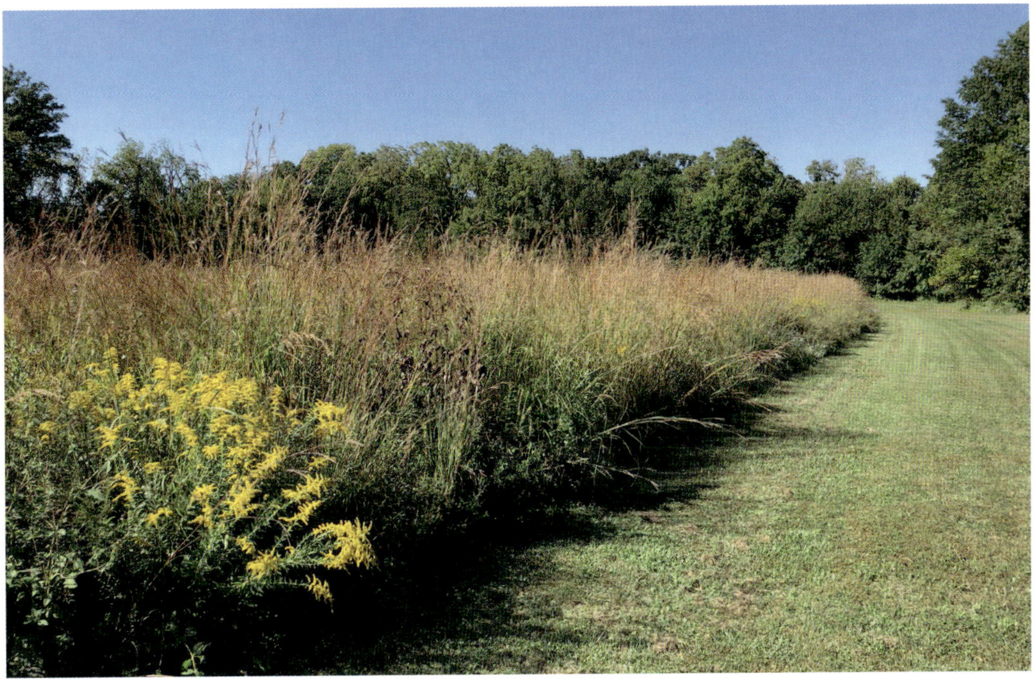

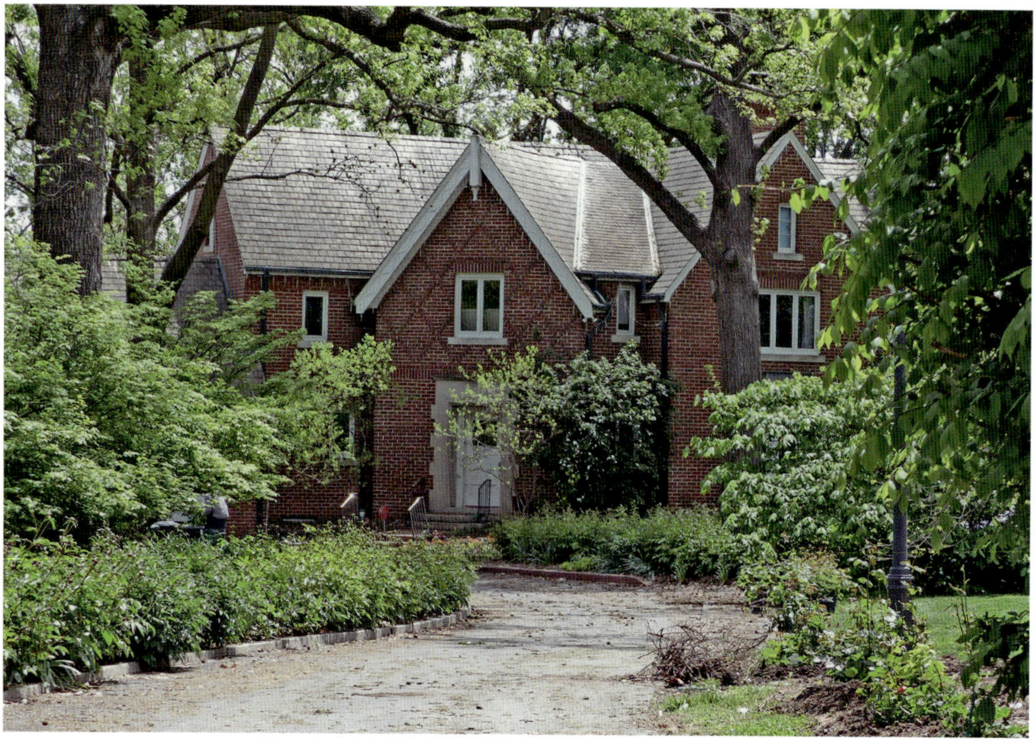

Two examples of Tudor Revival construction in brick, both predating World War II. [*Courtesy of Robert Mazrim*]

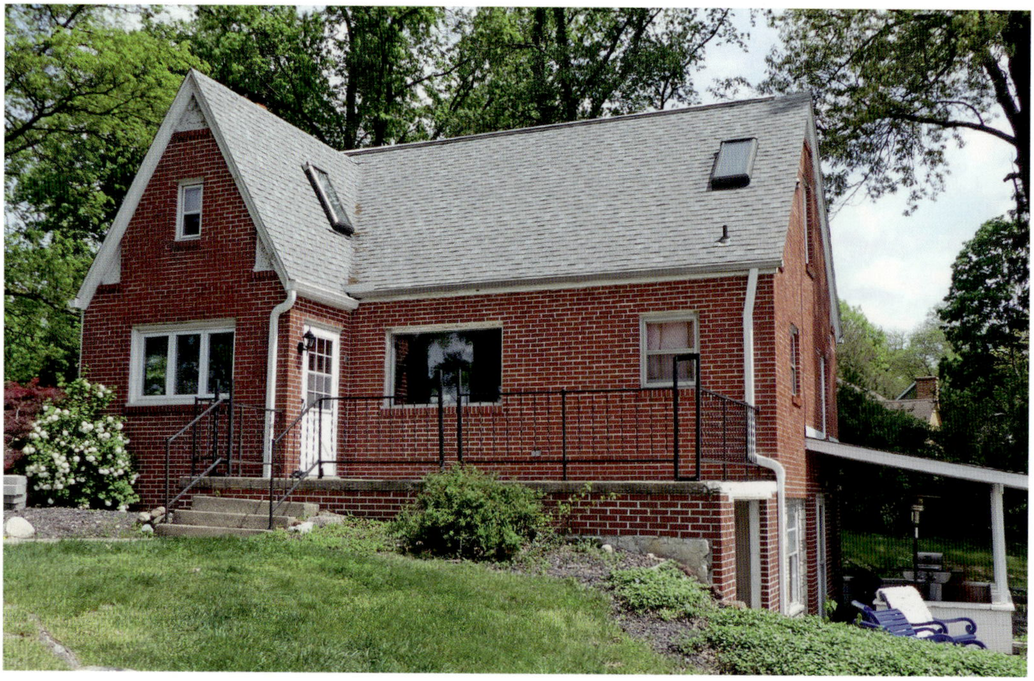

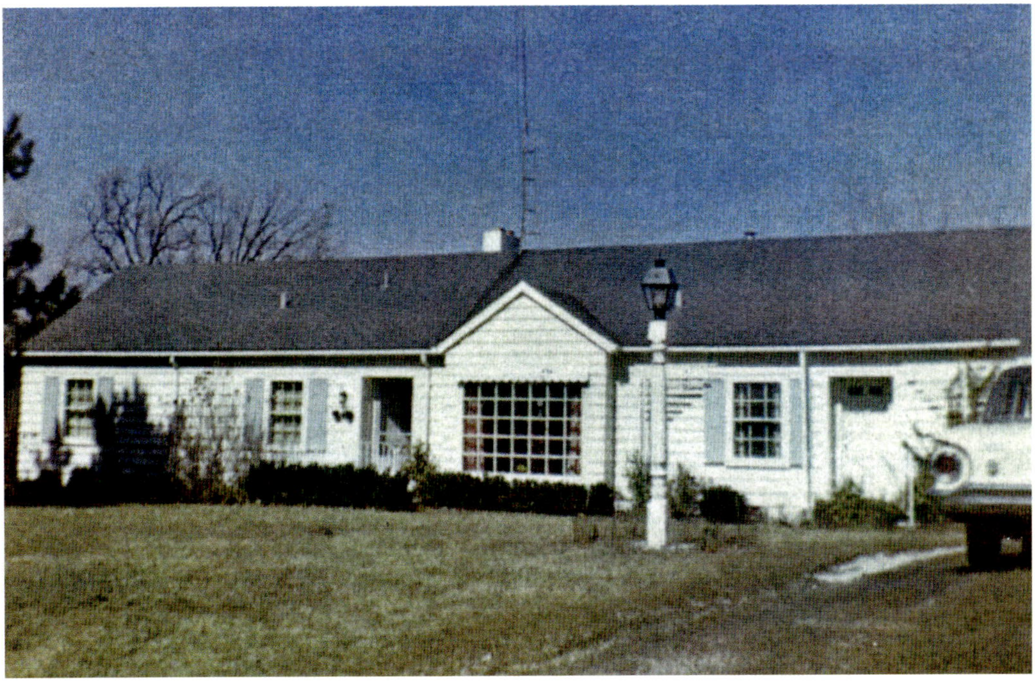

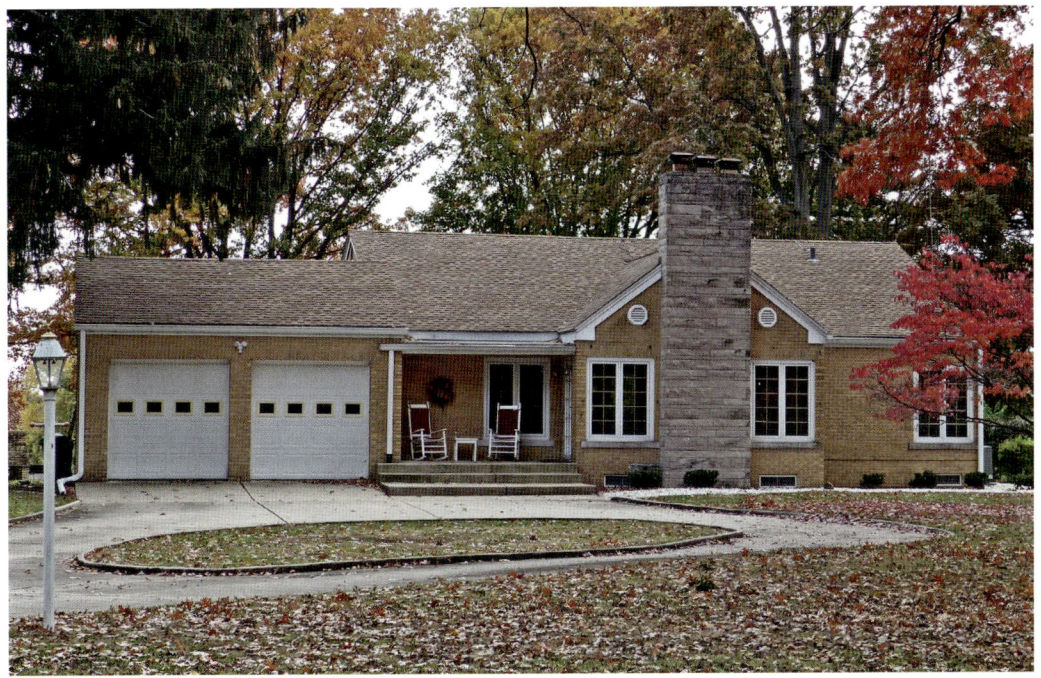

Mid-century ranch styles were quite common along the lake, as they were in a variety of new neighborhoods during the period. Often, they included architectural details that referred to older, historical styles, including the Cape Cod touches to the ranch house constructed around 1950 in Orchard Lane (above). [*Courtesy of Robert Mazrim*]

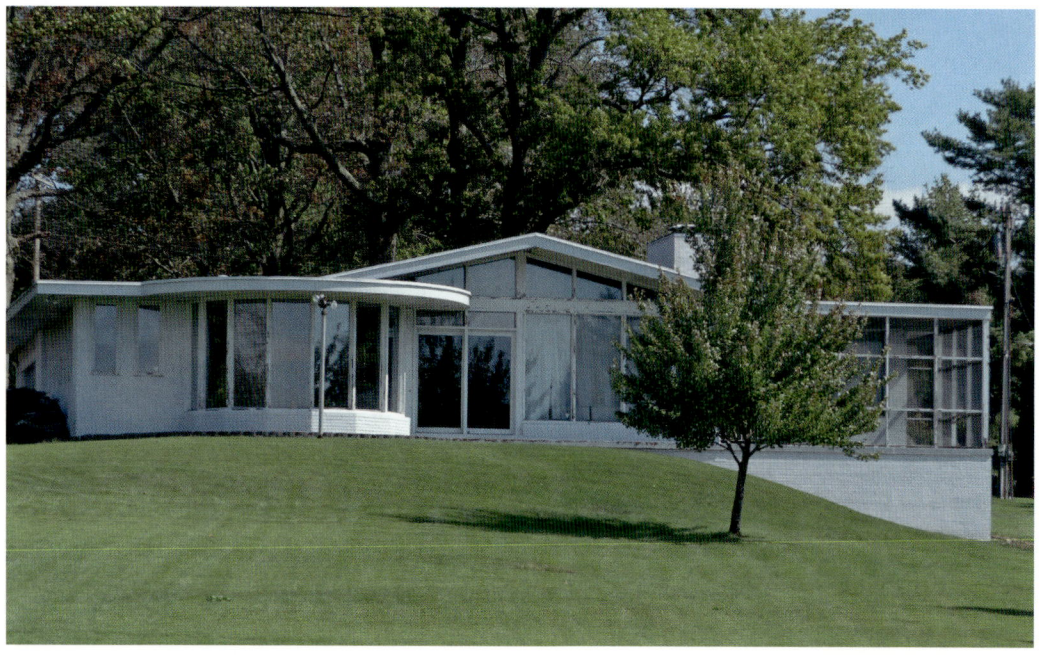

After the late 1950s, new styles appeared on the lake, including Contemporary or International styles, such as those shown here from East Shore Lane and East Lake Drive. These modernist styles are rapidly being recognized for their contributions to the history of American architecture, and are now old enough to be considered for inclusion in the National Register of Historic Places. [*Courtesy of Robert Mazrim*]

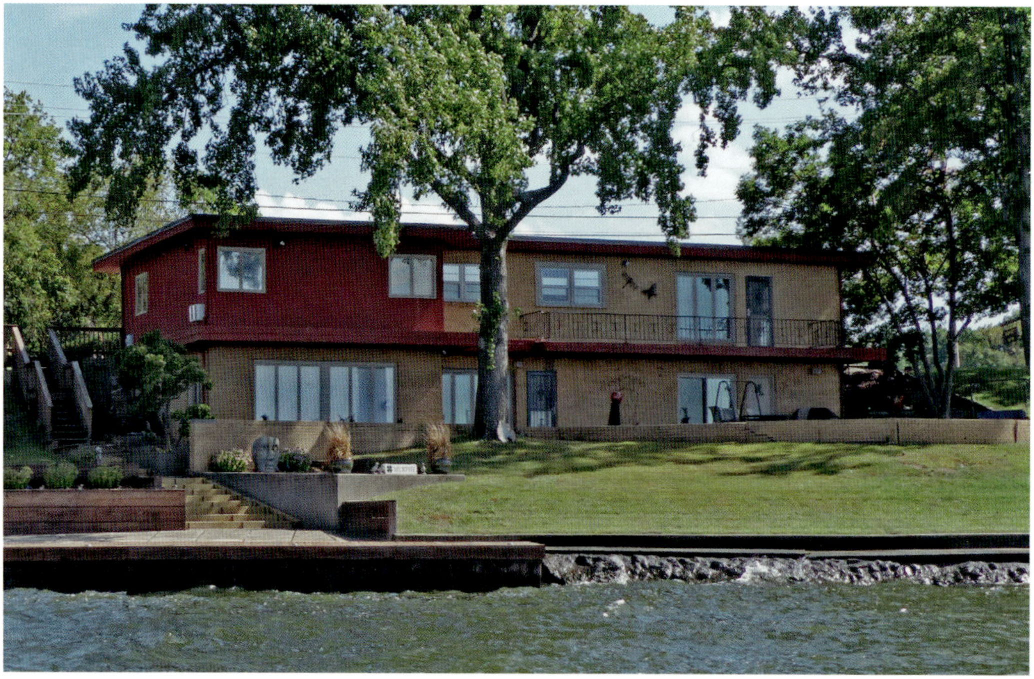

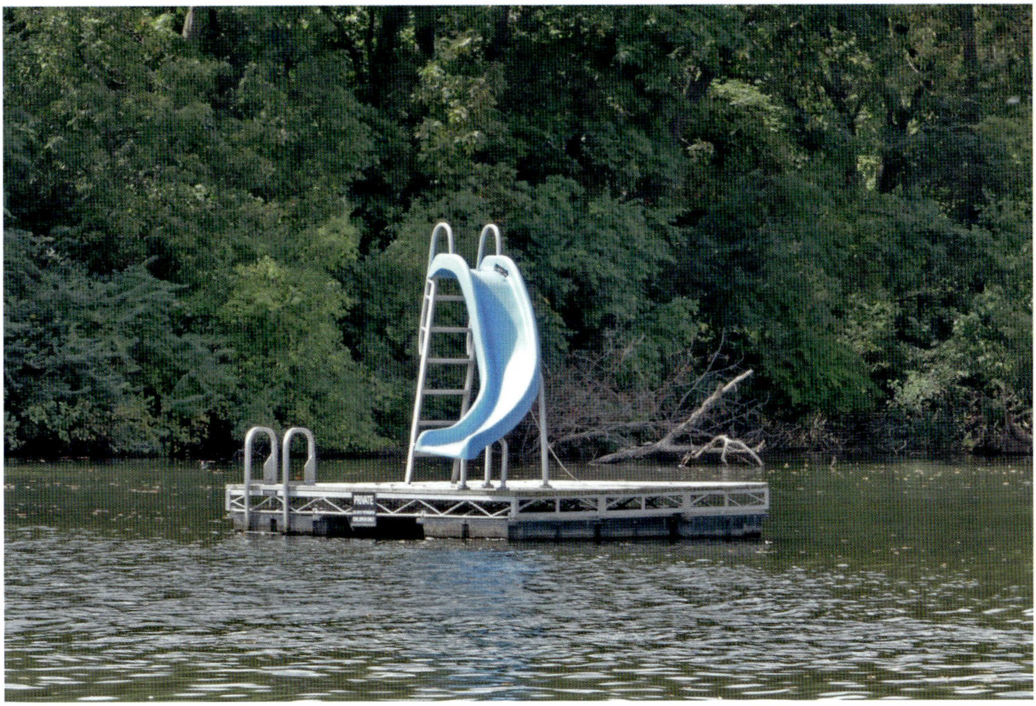

As they were lakeshore dwellings, many of these properties also included recreational amenities such as boathouses, shoreline patios and barbecue pits, beaches, and offshore floating docks. [*Courtesy of Robert Mazrim*]

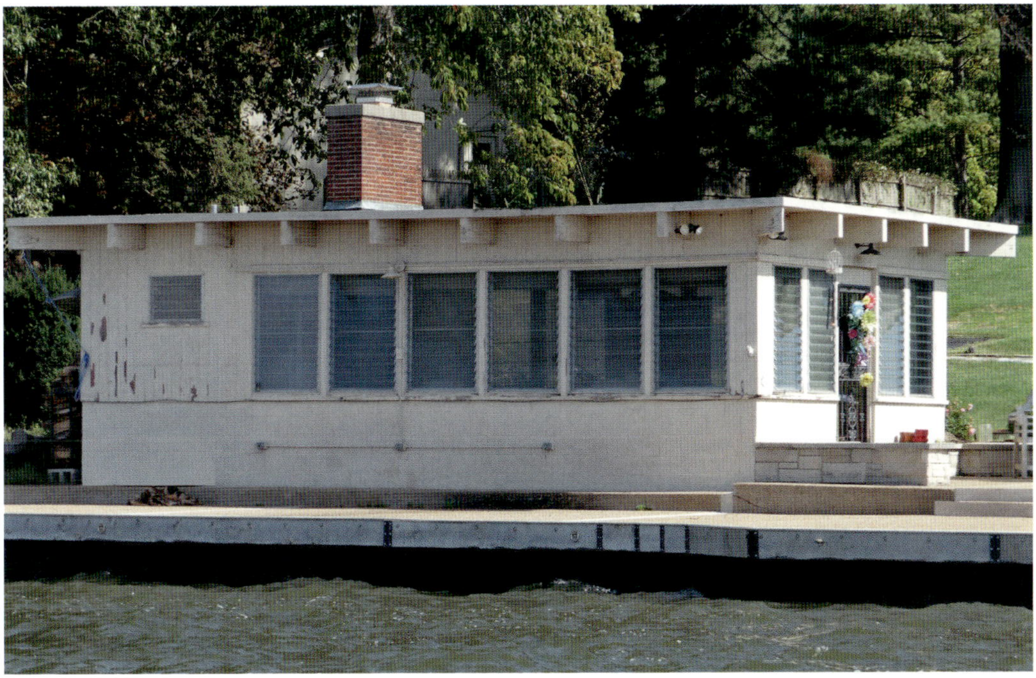

5

Things to Do

The vision for Lake Springfield included much more than a source of drinking water and power for the city. Aside from a well-planned concept for residential development, a series of municipal attractions were implemented into the overall design that would result in a vibrant recreational community surrounding the reservoir. Boat launches, a series of parks, two beaches and beach houses, provisions for numerous private clubs, scouting camps for children, a memorial garden, and a nature preserve created a unique community at the southern outskirts of the growing capitol city.

On the east side of the lake at Lindsay Bridge was a "recreation center," consisting of park grounds, improved picnic areas, playground equipment, ball diamonds, horseshoe courts, boat docks and launching ramps, and a large beach house and public beach. Today this area is known as Center, Lake, and Beach parks—three of many parks that now surround the water. The stone Lake Springfield Beach House was built in 1934. The beach itself, large enough to support 1,500 bathers, was protected by a breakwater that allowed the water along the beach to be separated from the main body of the lake. The swimming area was actually chlorinated behind the breakwater. Unfortunately, segregation was still practiced in Springfield during the 1940s, and this reached the new lake community. On the west side of the lake, a smaller beach and beach house ("Bridge View") was opened in 1936 for African American residents of the city, who were not allowed to bathe at the city beach. In 1952, the Lake Springfield Beach was desegregated.

A founding member of the Springfield Civic Garden Club, Harriet Knudson, persuaded the Garden Clubs of Illinois to oversee the creation of a designed park along the water featuring the native flora of central Illinois. Jens Jensen, an internationally-known landscape architect, designed what was intended as a "living tribute" to Abraham Lincoln, native son of the Springfield area. Plans were drawn in 1935, and Jensen himself selected a site along the eastern shore from four potential locations. In 1936, a 63-acre site (then consisting of farm fields and sparsely wooded areas) began its slow transformation into "Lincoln Memorial Gardens." Today, the gardens feature walking trails through mature native hardwoods, forest wildflowers, and restored prairie grass.

Springfield Park District trustee Henson C. Robinson conceived the idea of a children's zoo for the city. A 10+ acre site near the lake and beach house was donated by the City of Springfield in 1967 and the Henson C. Robinson Children's Zoo was opened to the public in June 1970.

Construction of an eighteen-hole, public golf course on the northeast shore of the lake began in 1948, initially through the efforts of the Junior Chamber of Commerce and the Allis Chalmers Company. The 200-acre site, first named the Lakeview Public Golf Course, was initiated as a private, non-profit enterprise. Work on the course lagged for many years, however, and the City of Springfield agreed to take control of the course in 1956. In 1957, the course's name was changed to Lincoln Greens.

Another recreational venue was established near the water in 1950, with the creation of the Springfield Municipal Opera Association. The association, better known as the Muni Opera, leased a 55-acre site from the city of Springfield near the east shore of the lake. A stage and other facilities were soon constructed. The organization produced several seasons of performances before lack of financial support forced the closure of the opera in 1956. Renewed public interest in the 1960s led to a successful fundraising campaign, and in 1965 the stage was rebuilt, and the opera reopened.

Of course, fishing has always been a popular pastime at the lake. Two rearing ponds for fishes were located just east of the lake, where carp, catfish, crappie, black bass, white perch, bluegill, and sunfish were raised and introduced into the lake. Catfish, bass, and crappie remain the most popular sporting fish today. In 2019, a record setting 98-pound blue catfish was caught and released.

Several public boat launches were constructed at various points along the shoreline, allowing those who did not live on the water or belong to private clubs to launch their crafts. In the spring of 1991, the Lake Springfield Marina was established at the recently vacated Knights of Columbus club on the western shore of the lake. The small facility provided boat slip and boat rentals, and also included a restaurant. In 2001, the marina relocated to a site further up the former Sugar Creek valley, along a section of abandoned Route 66. Today, the much-expanded marina continues to provide boat slip rentals, boat sales and rentals, fuel, and repairs. It is also known for its Fourth of July fireworks displays.

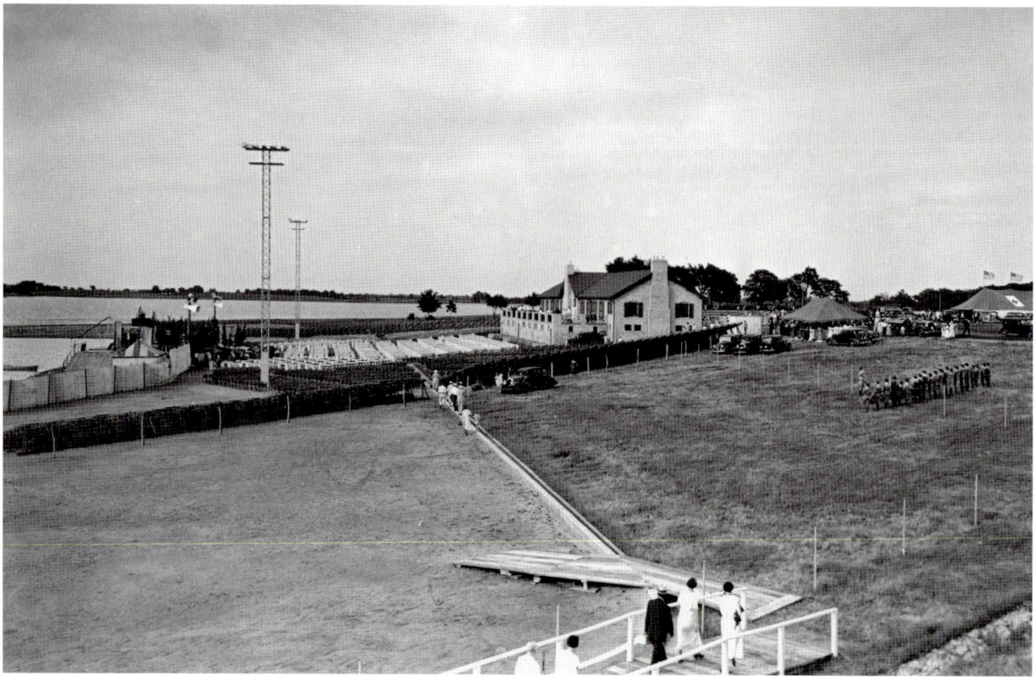

A number of municipal recreational facilities were planned as part of the original construction of the lake in the 1930s. Shown here is the dedication of the public beach house, located near Lindsay Bridge, along with a very busy day at the beach during the 1950s. [*Sangamon Valley Collection, Lincoln Library*]

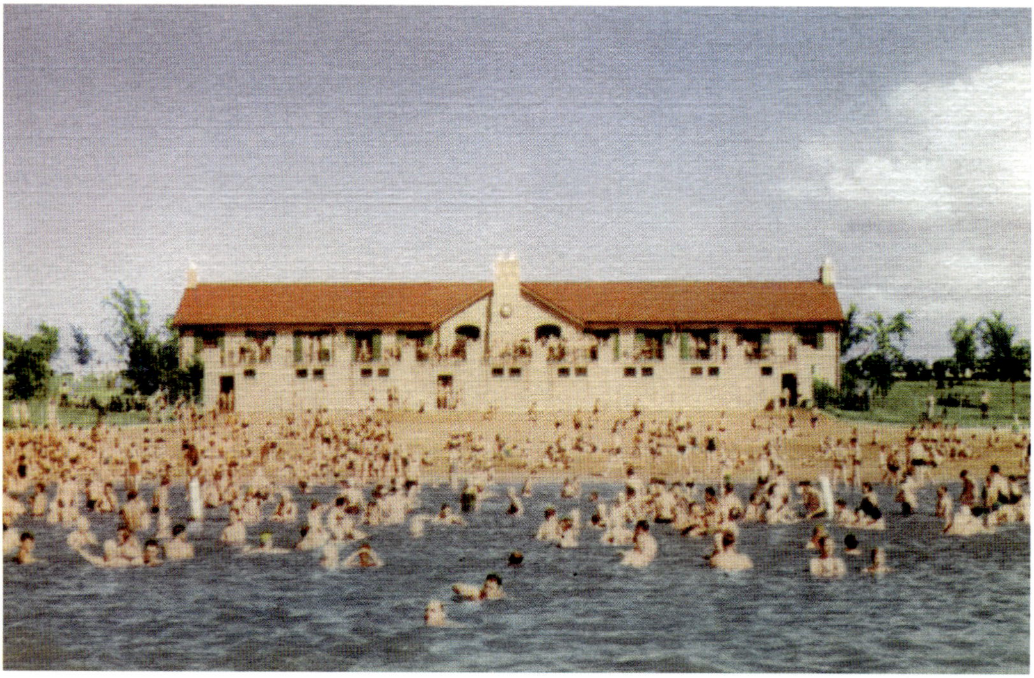

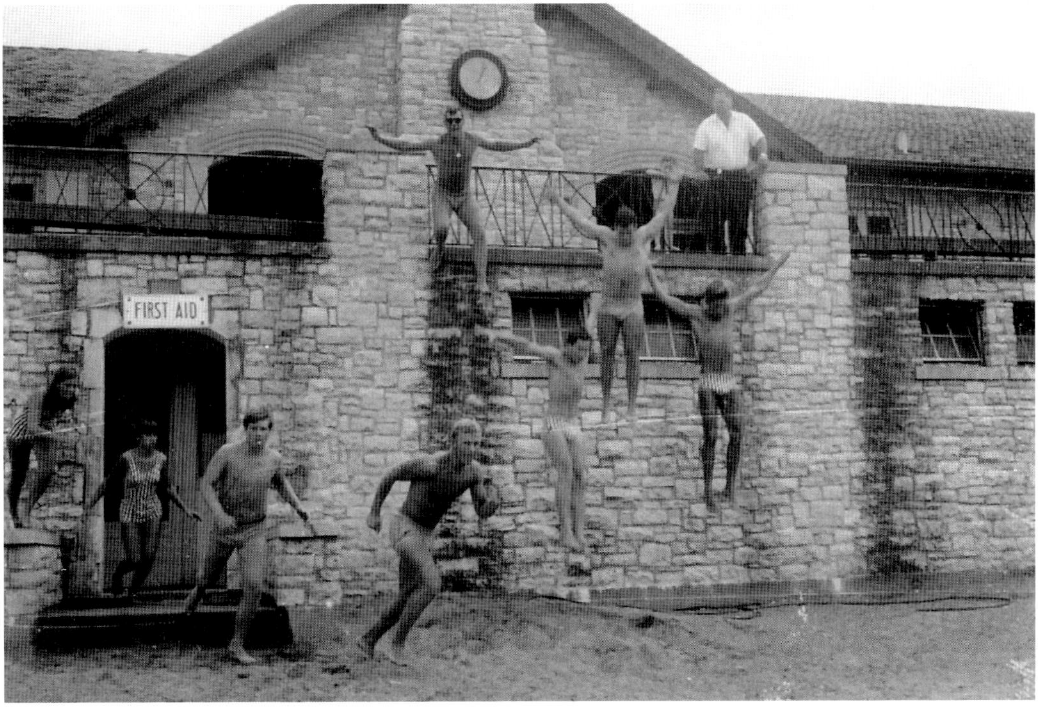

The beach house was busy during its 1940s-1960s heyday, and was equipped with a number of diving boards, slides, and a beach monitored by lifeguards. A sea wall enclosing the water around the beach even allowed for the chlorination of the lake water adjacent to the beach. [*Sangamon Valley Collection, Lincoln Library*]

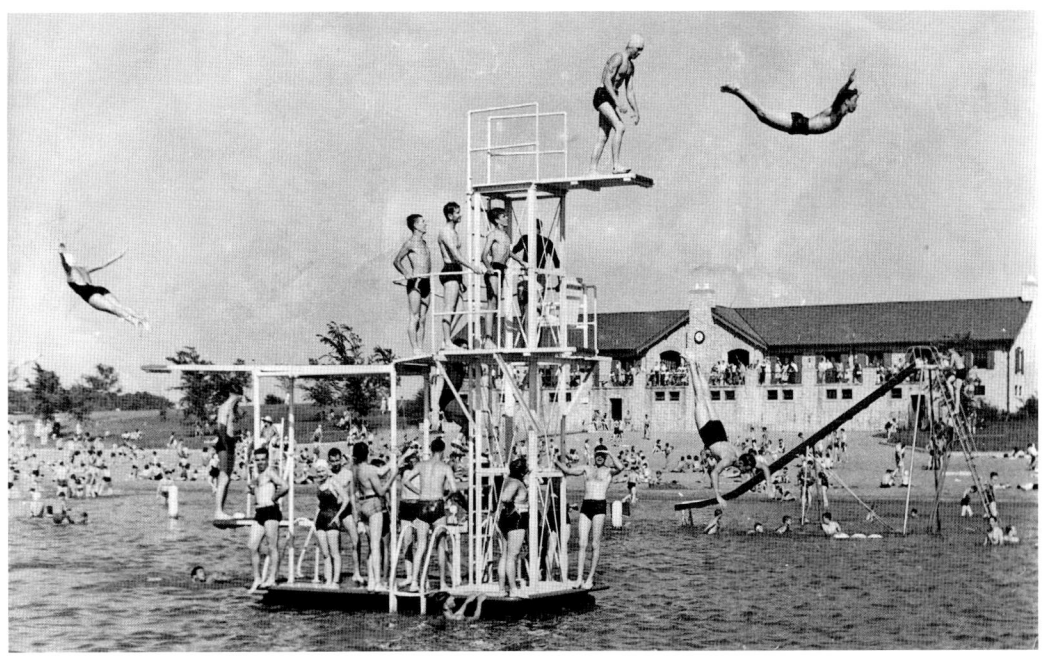

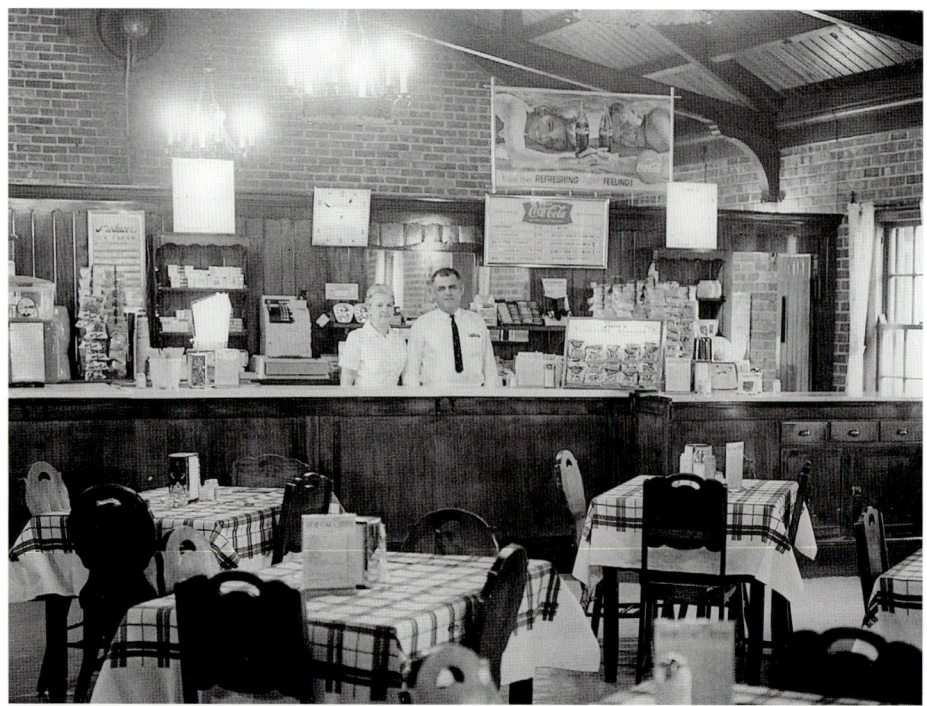

For a time, the beach house was also equipped with an indoor snack bar (above). [*Sangamon Valley Collection, Lincoln Library*] Today, however, the beach is unvisited, except for dozens of seagulls seen here sunning themselves on the sand. The City of Springfield closed the beach house in 2008, and the future of this historic facility is unclear. [*Courtesy of Robert Mazrim*]

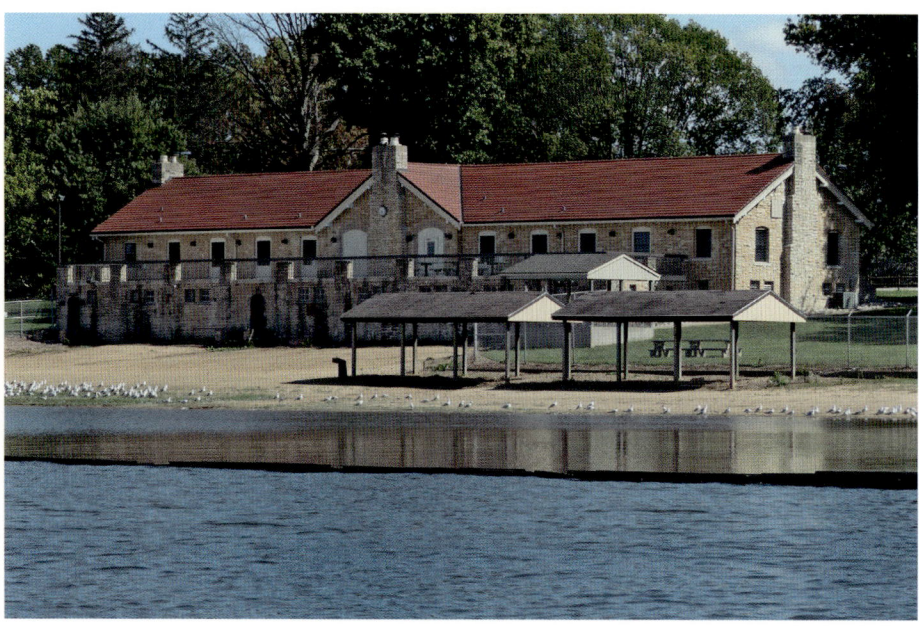

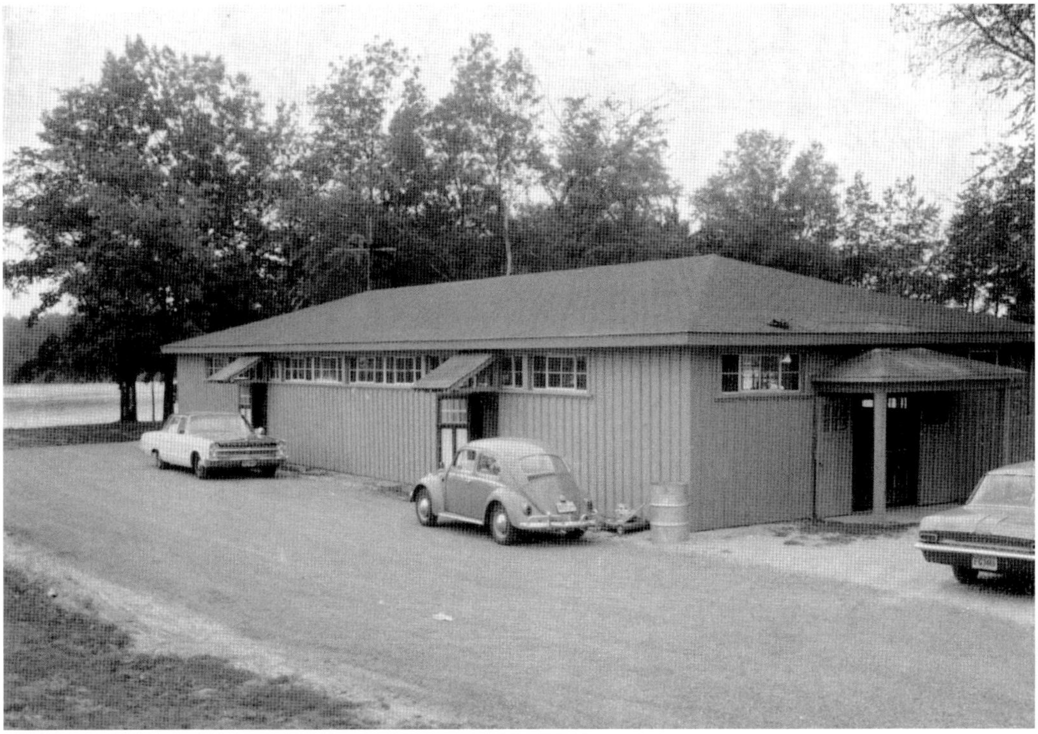

Unfortunately, like many at the time, the Lake Springfield community was often a segregated one. This was particularly evident at the beach facilities. A second, much less grand beach house, "Bridgeview Beach," was constructed at the opposite end of the lake for use of the African American community in Springfield. It too was staffed by experienced lifeguards, however. By the mid-1960s the stone beach house at Lindsay Bridge had been integrated and the beach at Bridgeview eventually fell into disuse. [*Sangamon Valley Collection, Lincoln Library*]

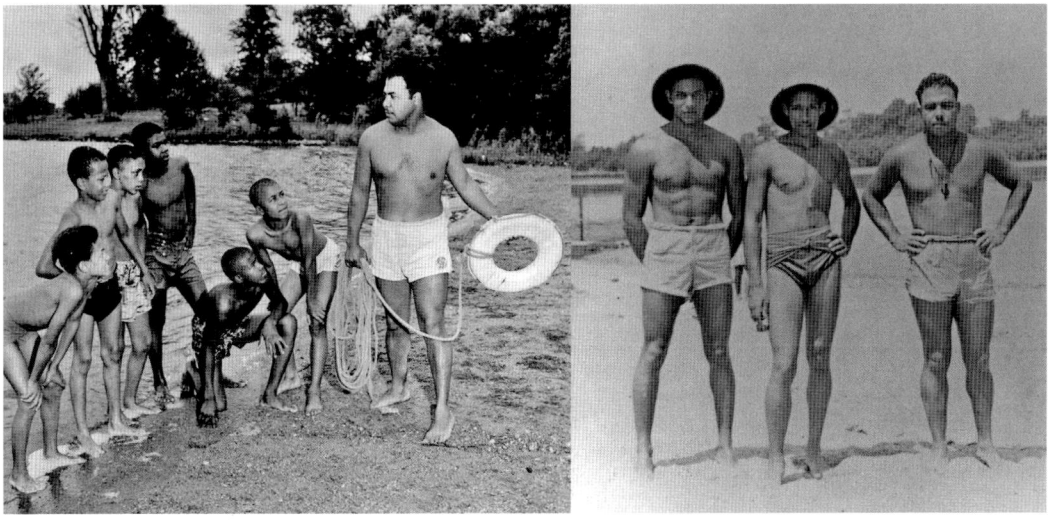

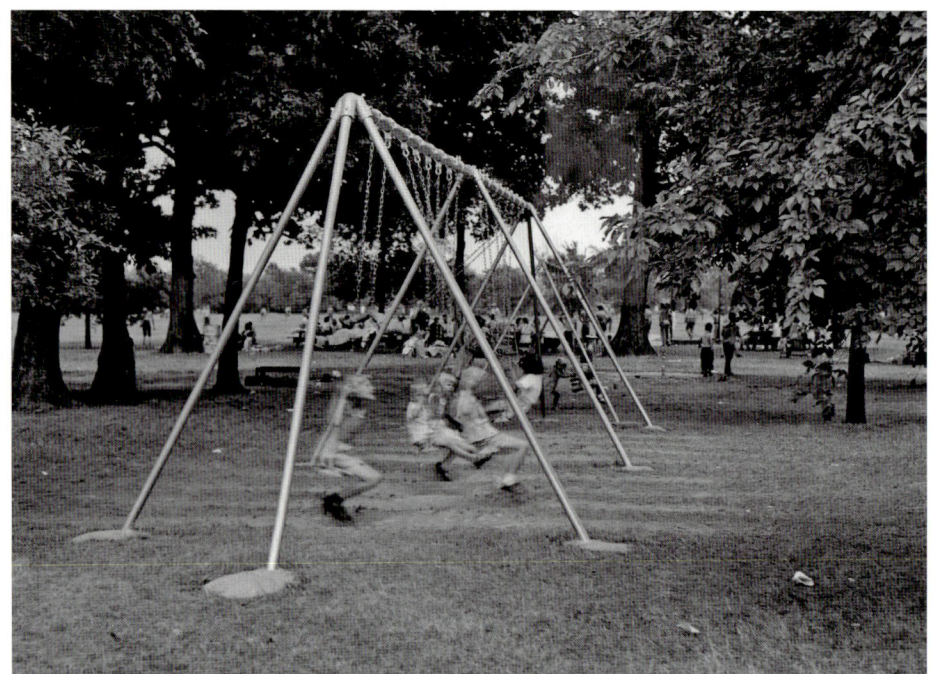

A total of eight city parks eventually surrounded the lake, providing playground equipment, picnic areas, and baseball diamonds. Most of these parks remain much like they were fifty years ago, although most of the vintage playground equipment has been updated in the last twenty years. [*Above: Sangamon Valley Collection, Lincoln Library; Below: Courtesy of Robert Mazrim*]

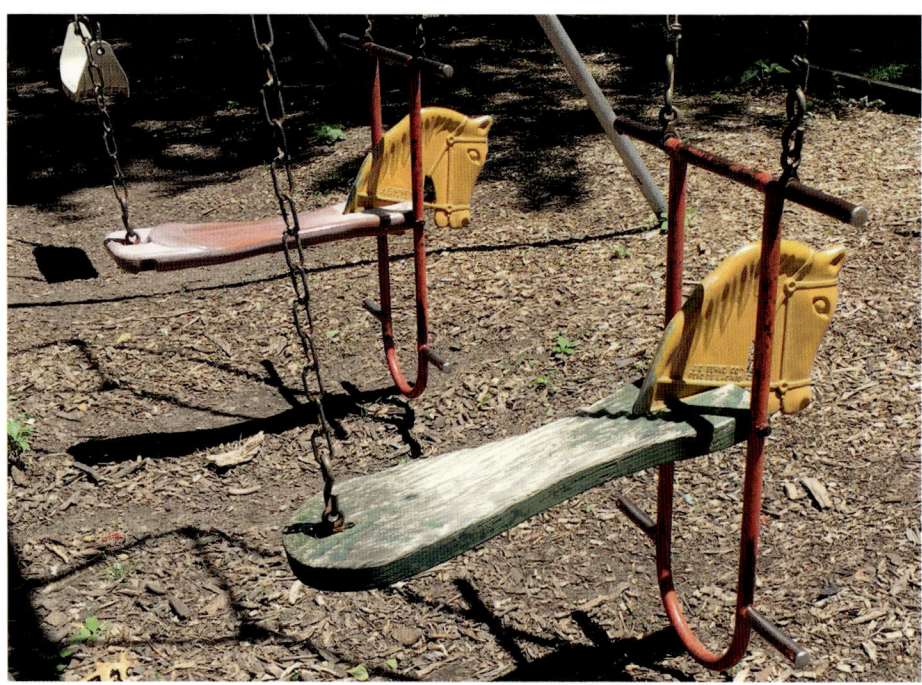

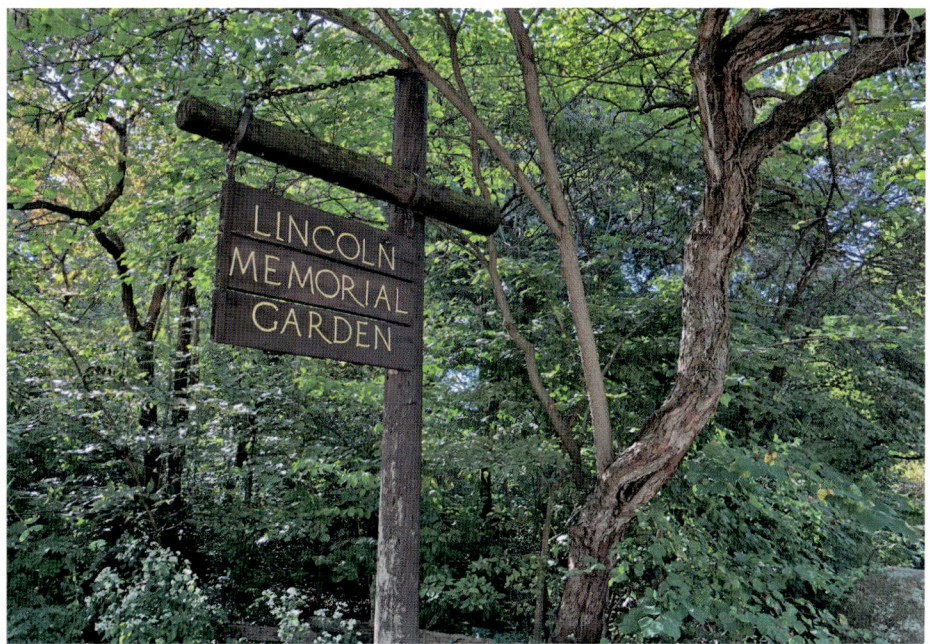

In 1935, plans were drawn for a "living tribute" to Abraham Lincoln consisting of a carefully designed 63-acre natural area planted in a variety of native plant species. Lincoln Memorial Gardens was designed by noted landscape artist Jens Jensen (middle), seen here with Harriet Knudsen (left) who spearheaded the effort, and Dr. T. J. Knudsen (right). [*Above: Courtesy of Robert Mazrim; Below: Sangamon Valley Collection, Lincoln Library*]

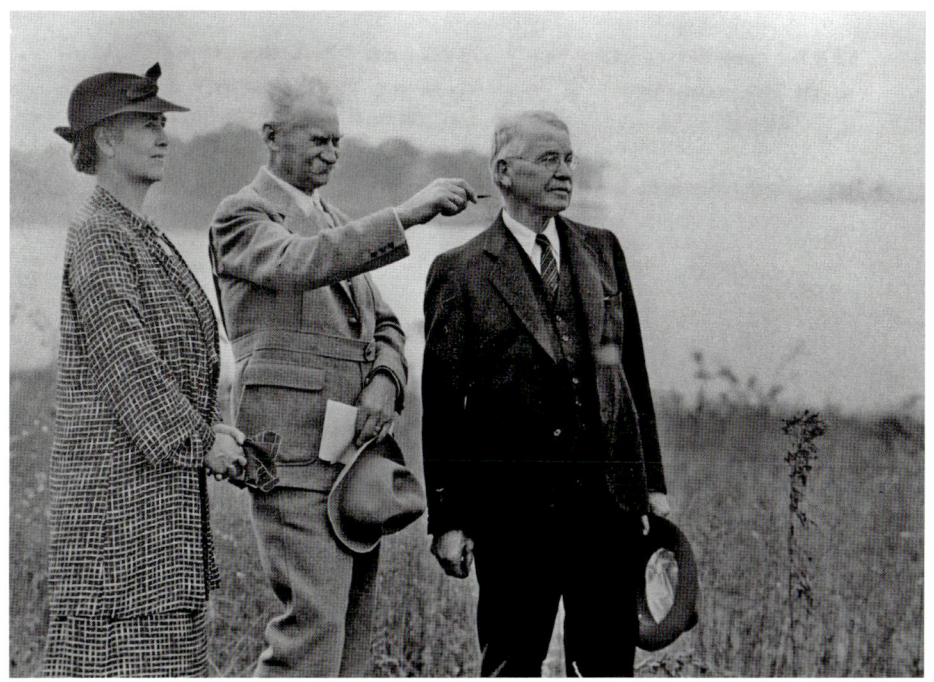

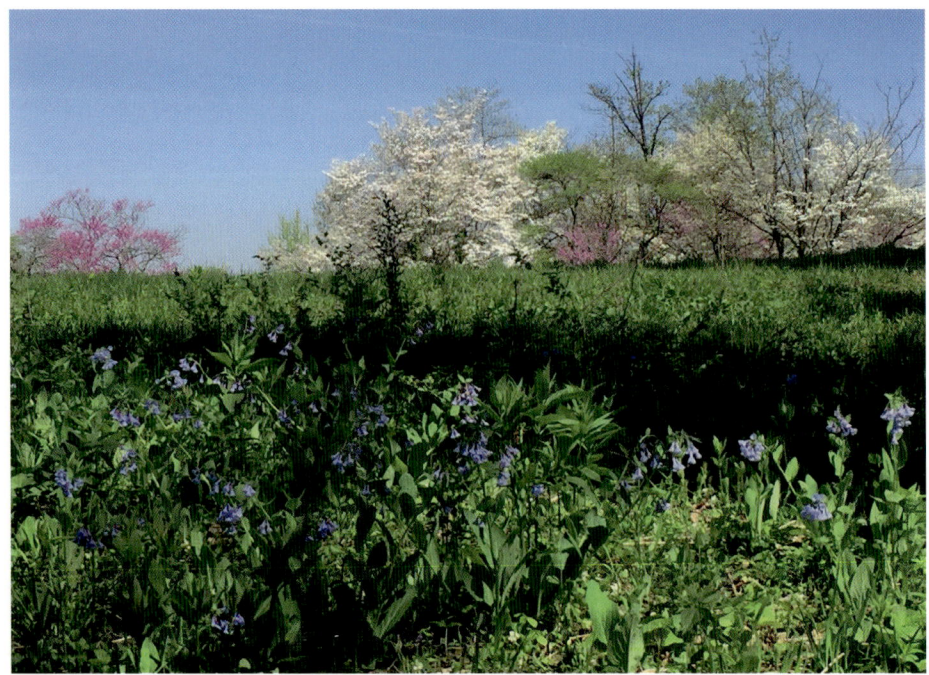

Today, the mature gardens feature, in the spring, native bluebells bracketed by dogwoods and redbuds. One of the eight original stone "council rings" that still punctuate the natural area is shown below. [*Courtesy of Robert Mazrim*]

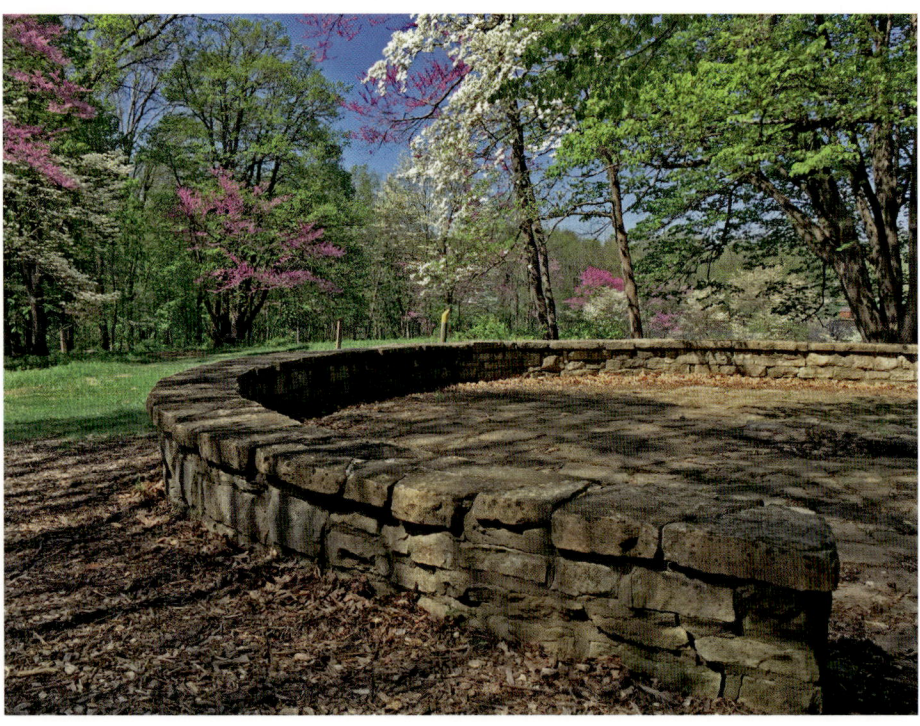

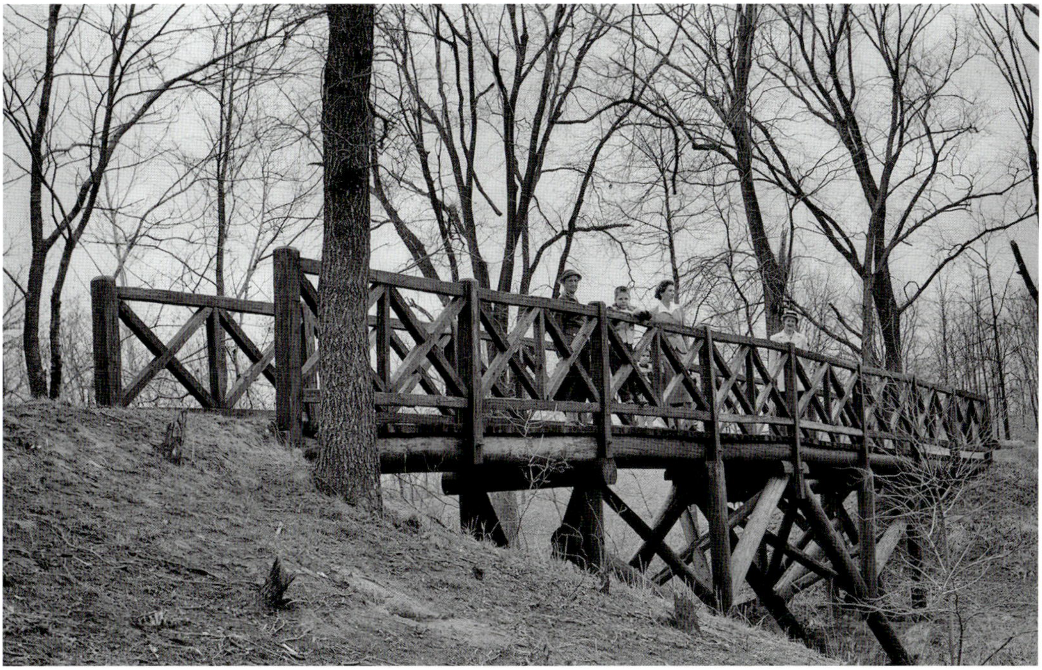

Things to Do

An early, primitive-style wooden bridge constructed over one of the small creeks that flow through Lincoln Memorial Gardens is shown above. In 1965, a visitor center, in a modernist style, was constructed to welcome guests to the natural area. [*Sangamon Valley Collection, Lincoln Library*]

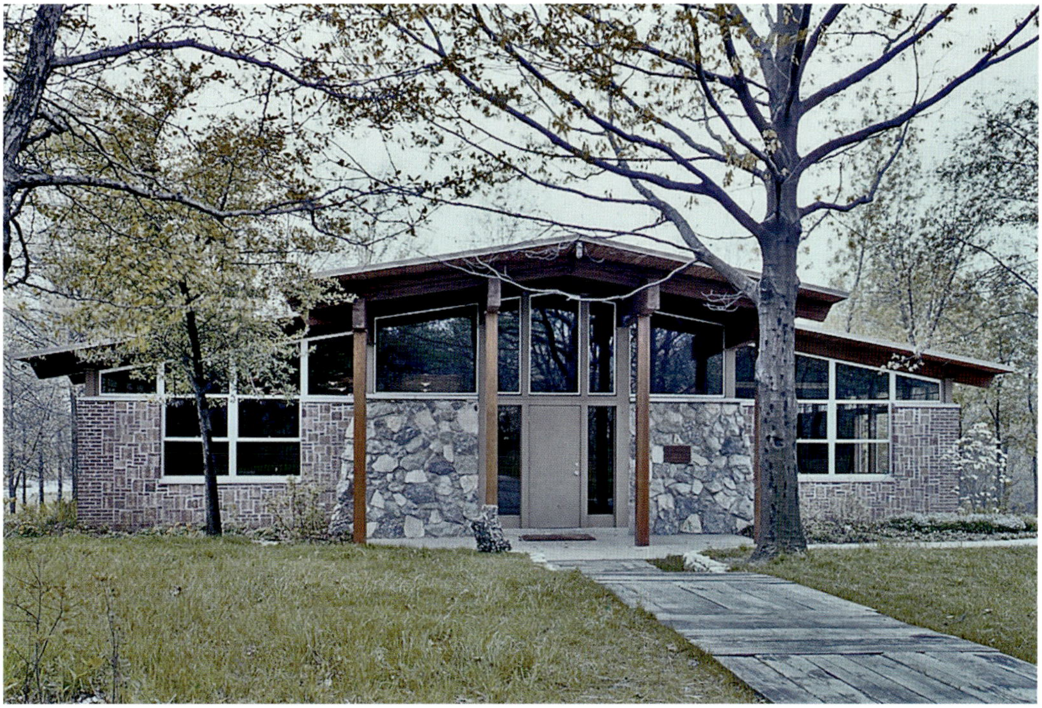

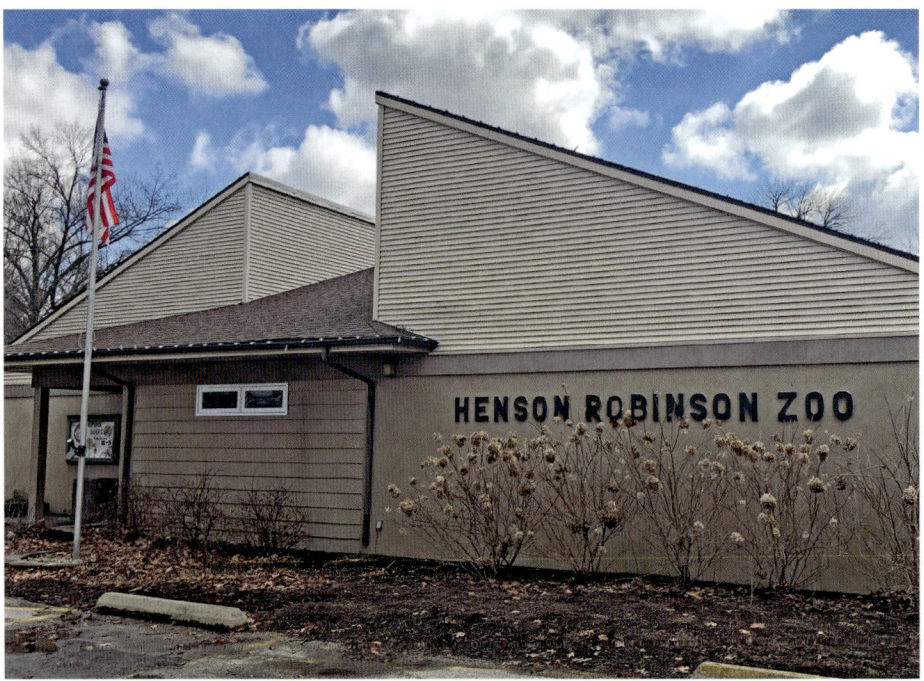

During the 1960s, the Springfield Park District began construction of a small zoo in a park area near the beach house. Henson Robinson Zoo was opened to the public in 1970, and among the dignitaries present at the opening was Marlon Perkins, popular star of the *Wild Kingdom* television program. [*Above: Courtesy of Robert Mazrim; Below: Sangamon Valley Collection, Lincoln Library*]

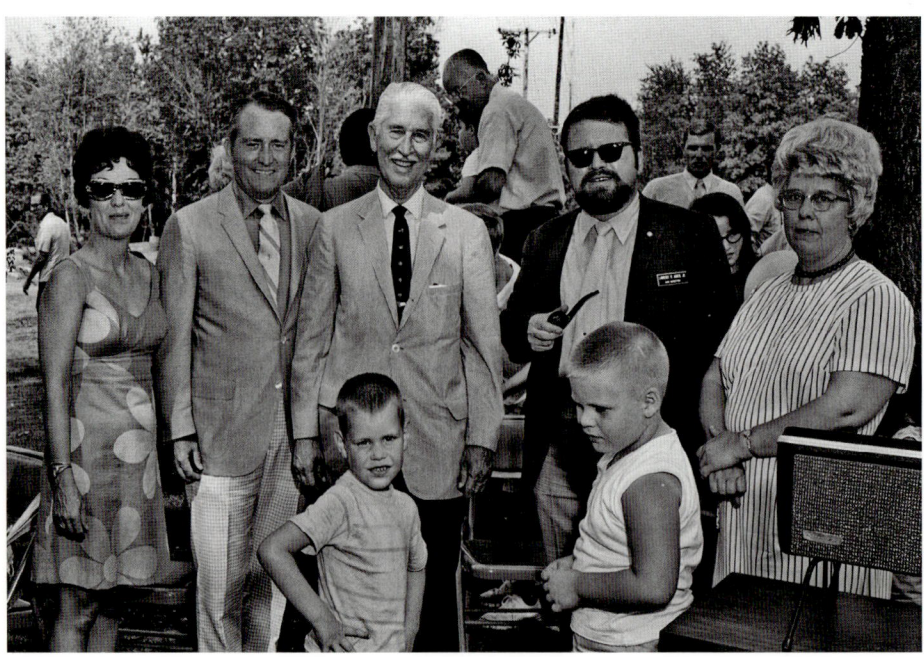

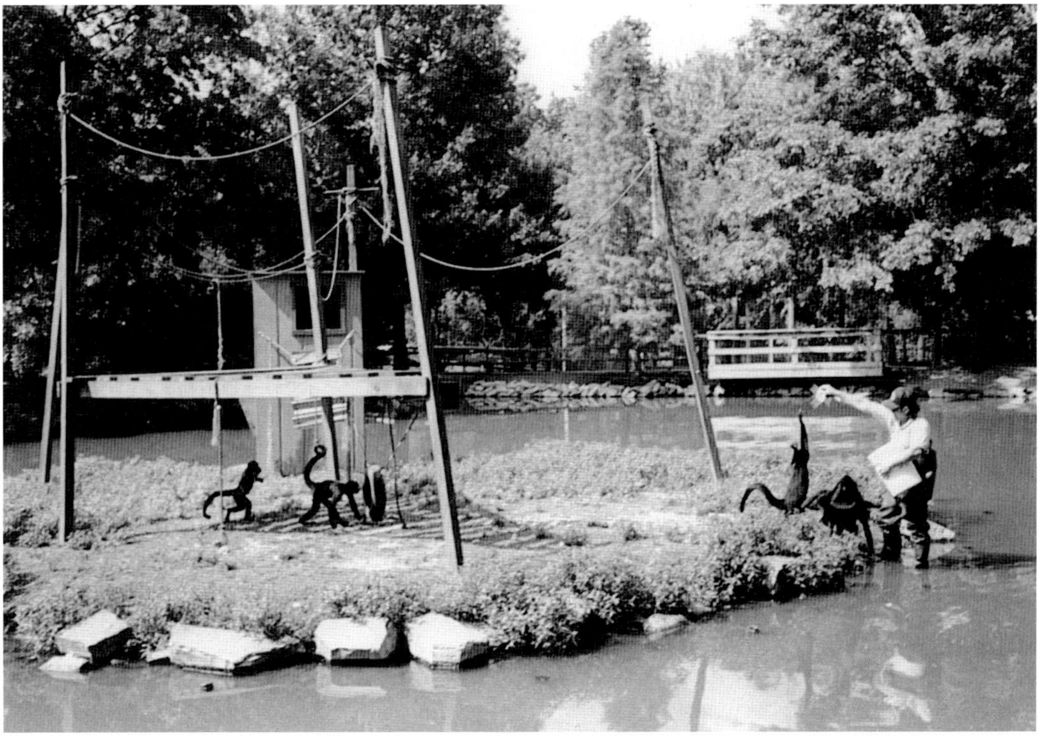

The habitats and residents of the zoo have changed somewhat over time. Here, spider monkeys are fed on a small island preserve, known as "Monkey Island" in a lagoon inside the zoo. Below is the ever-popular penguin habitat. [*Sangamon Valley Collection, Lincoln Library*]

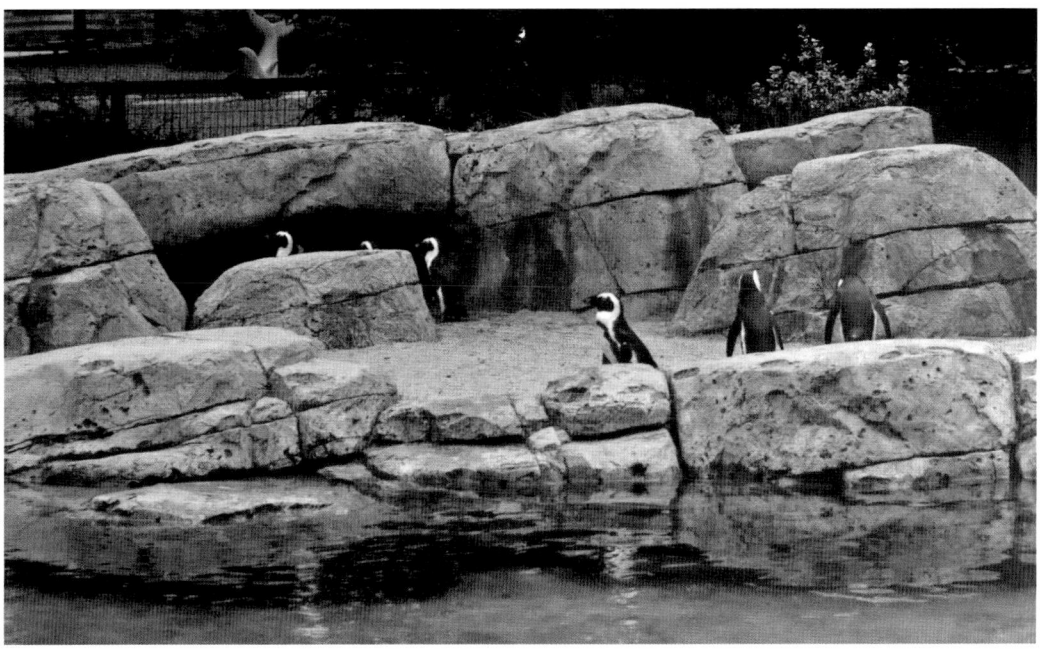

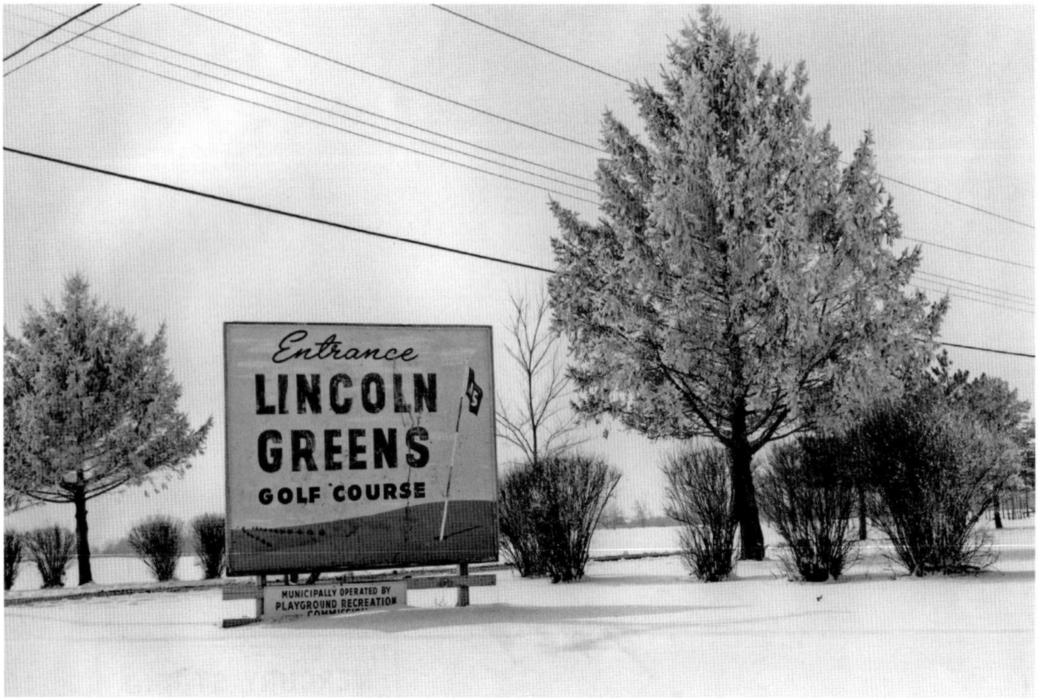

Lake Springfield in Illinois

In 1948, through the efforts of the Junior Chamber of Commerce and the Allis-Chalmers Company, an eighteen-hole public golf course was constructed along the northeast shore of the lake. It remains a popular course in the Springfield area today. [*Sangamon Valley Collection, Lincoln Library*]

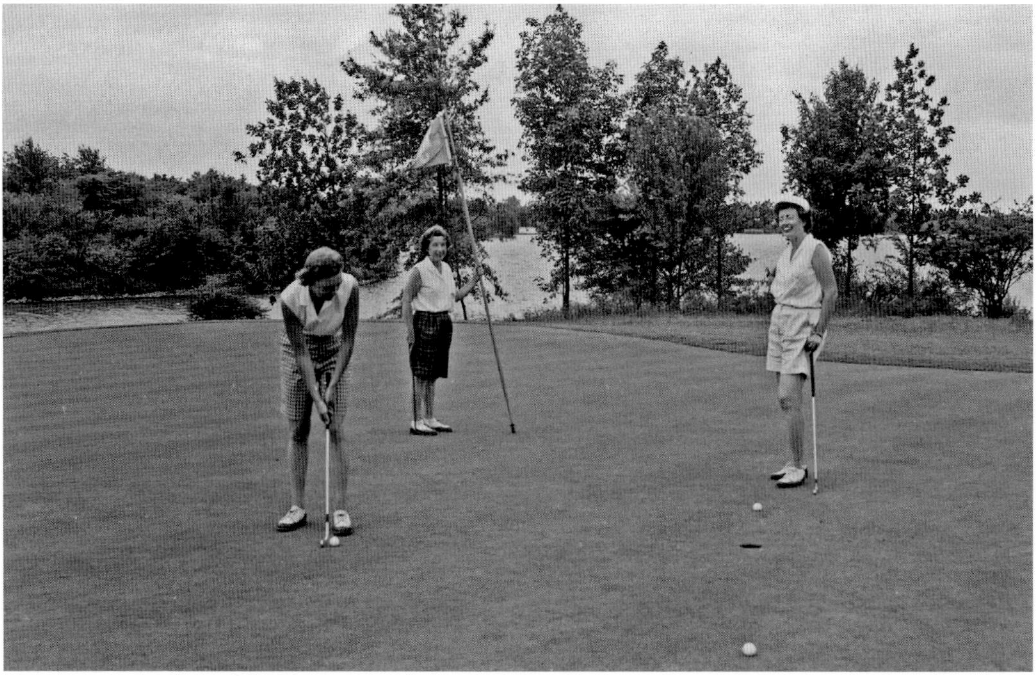

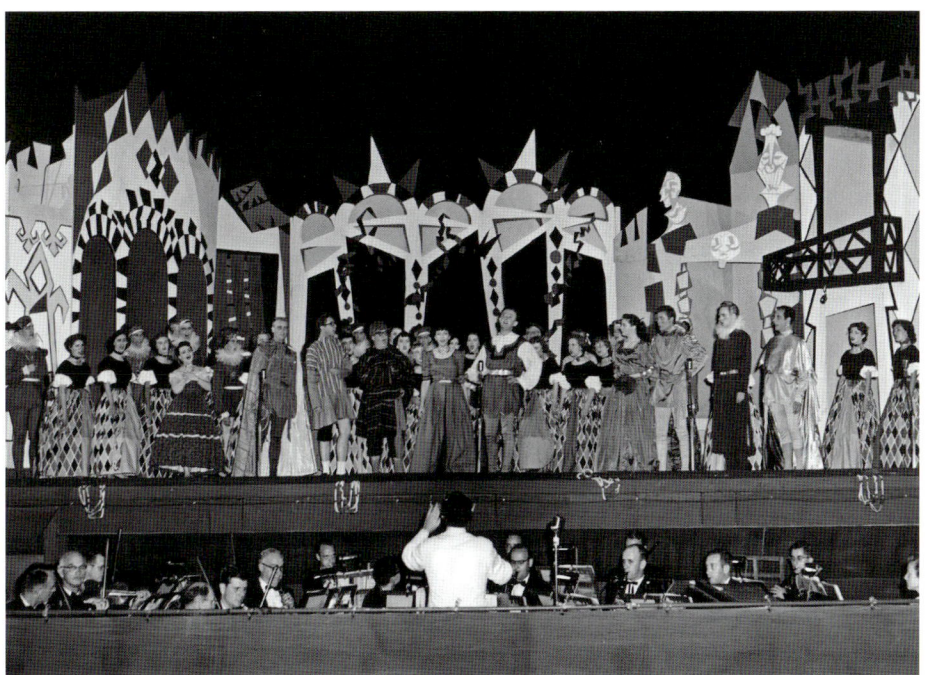

Just across East Lake Shore Drive from the lake, the Springfield Muni Opera was constructed in 1950. The facility features an outdoor stage situated in a secluded, natural environment. Shown here is the 1955 production of "Kiss Me Kate," along with a publicity photo for the 1993 production of "The Pirates of Penzance." [*Sangamon Valley Collection, Lincoln Library*]

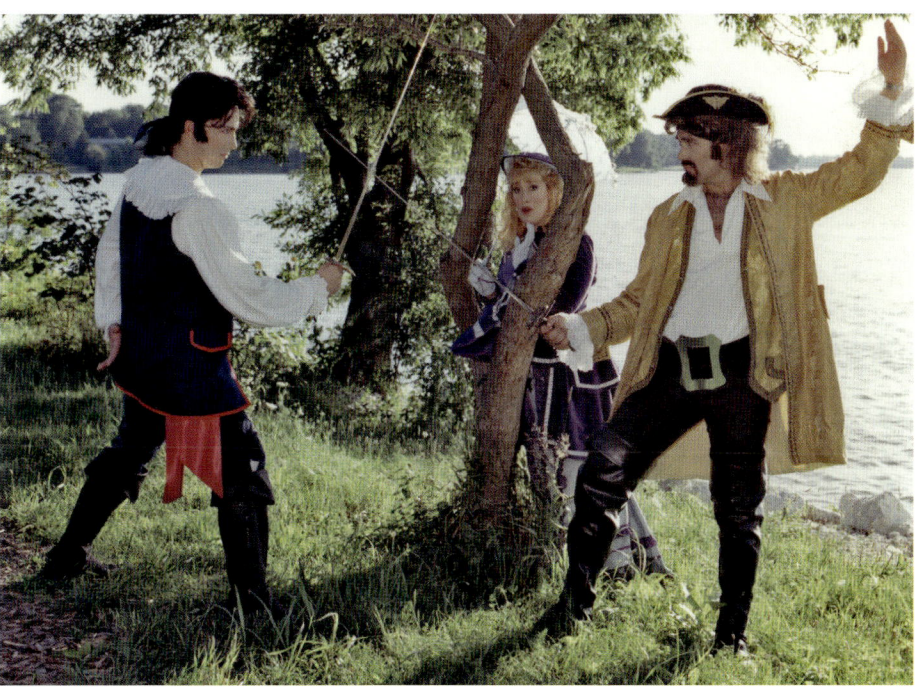

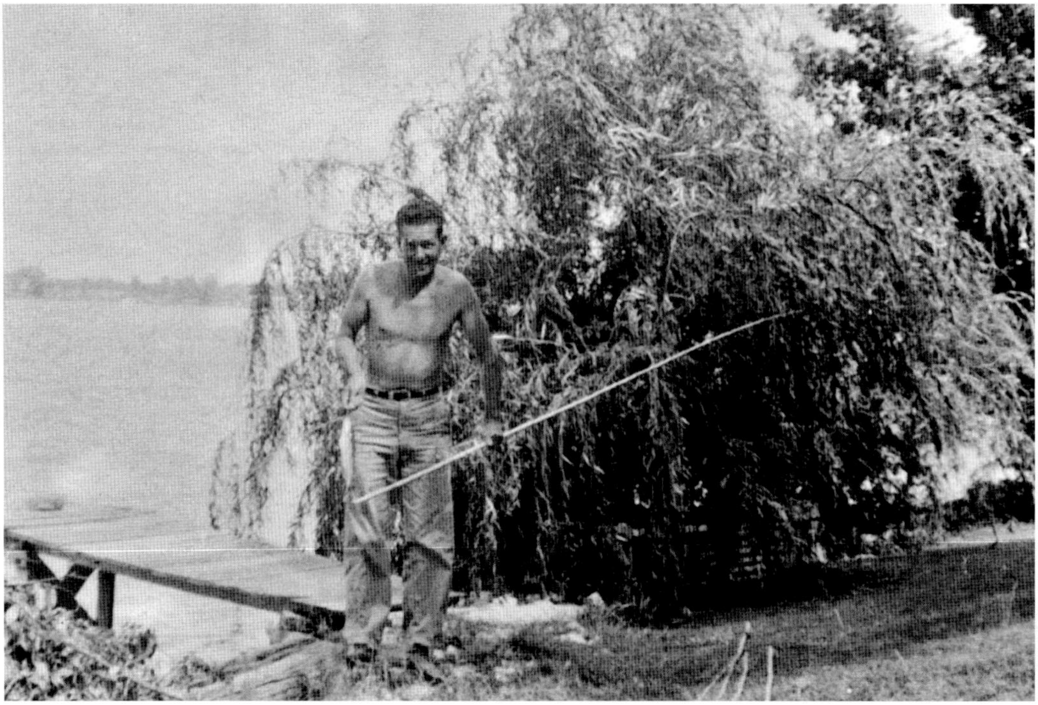

Of course, fishing has and always will be a popular pastime on the lake. Above, a bass is caught using a bamboo rod by a resident of East Hazel Dell Lane in 1949. [*Courtesy of Robert Mazrim*] Below, visitors to the lake cast off the shore of one of the city parks, sometime during the 1950s. [*Sangamon Valley Collection, Lincoln Library*]

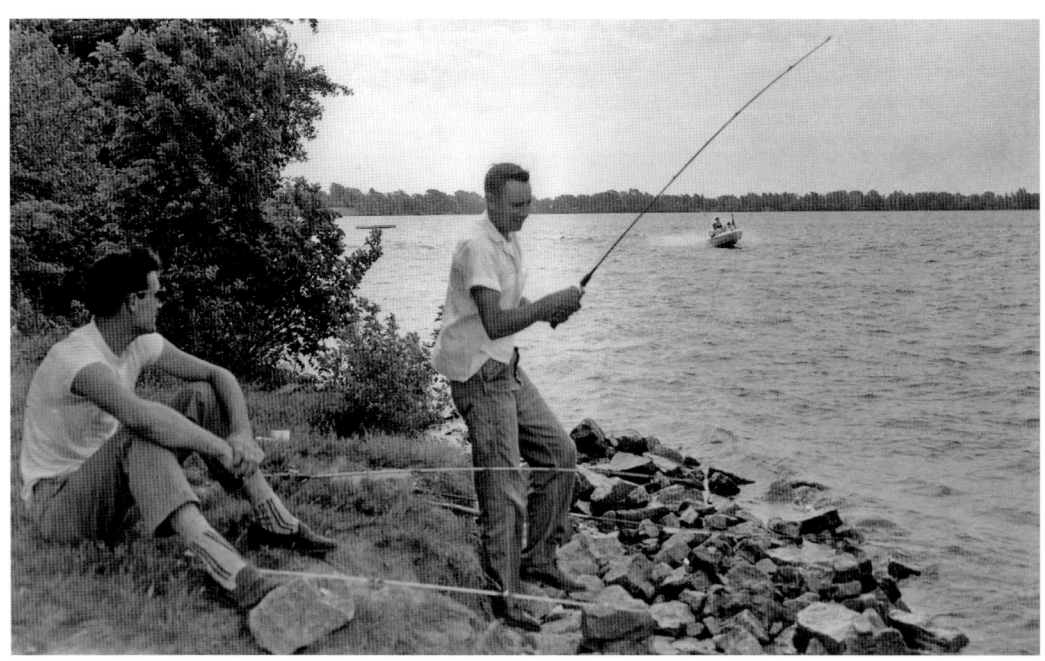

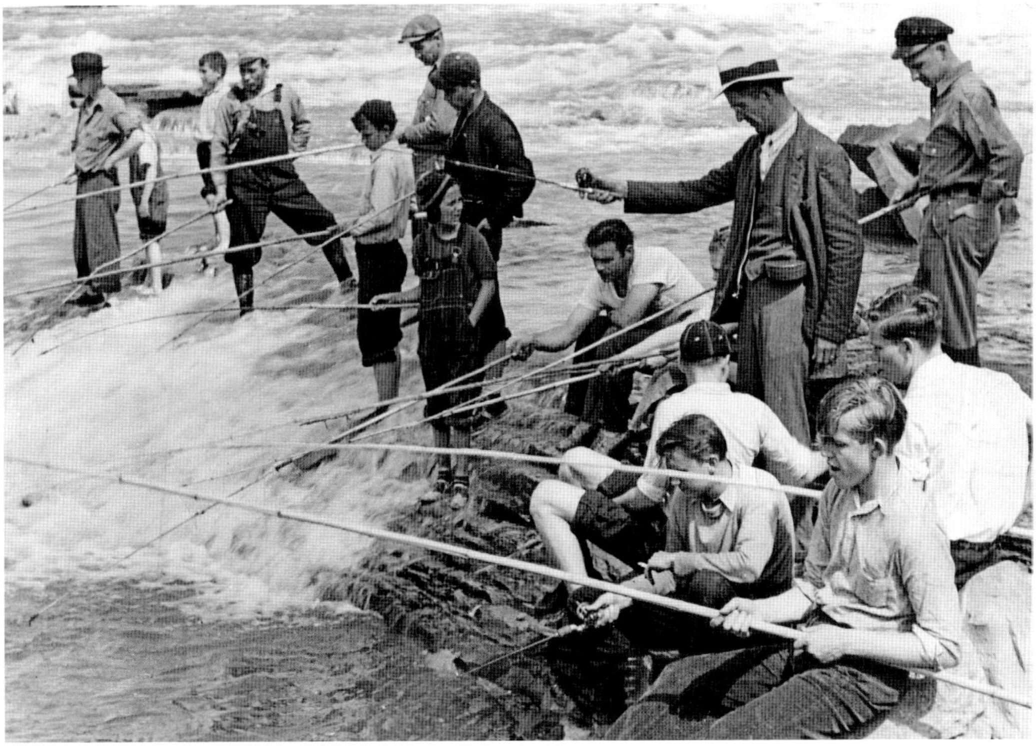

When one or more of the gates of the dam is open, water and fish alike are cast over its gates and across a rocky area composed of rapids and small pools. Here, a number of local residents crowd one of those pools, attempting to catch the bewildered fish after their tumble over the dam. Below, two gentlemen pose with a large catfish caught in Lake Springfield. [*Sangamon Valley Collection, Lincoln Library*]

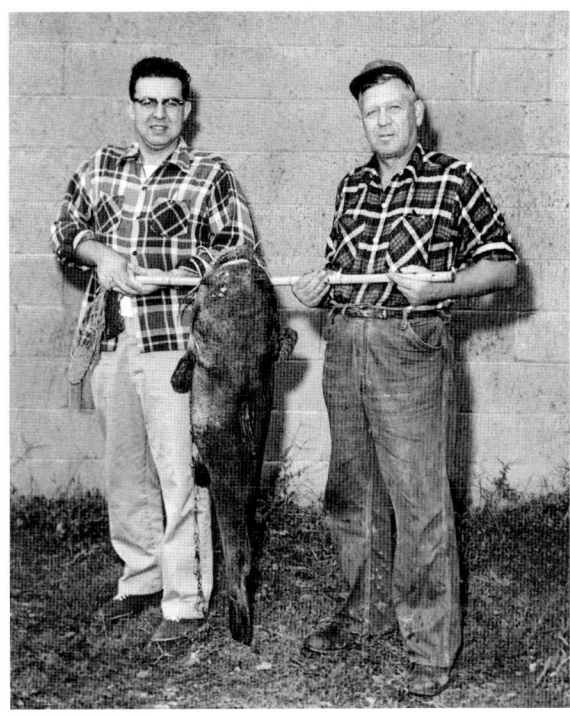

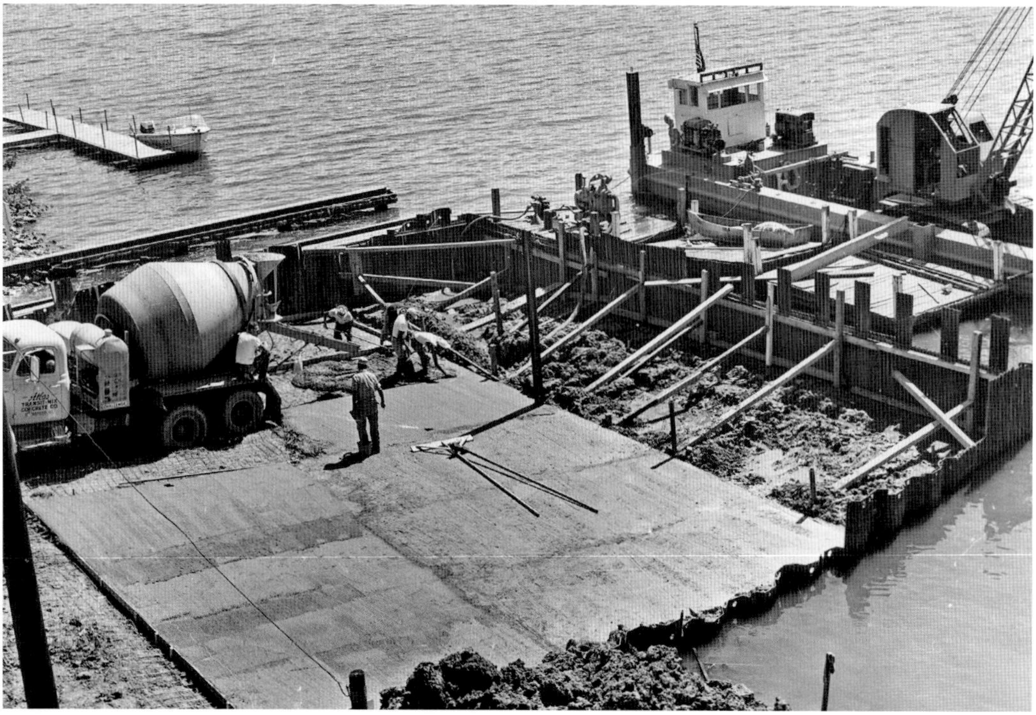

Boating is another important recreational activity on the lake. But before one can go boating, one must get the craft into the water. Several public boat ramps were constructed at various points along the lake shore. The ramp near Lindsay Bridge, below, remains one of the busiest today. [*Sangamon Valley Collection, Lincoln Library*]

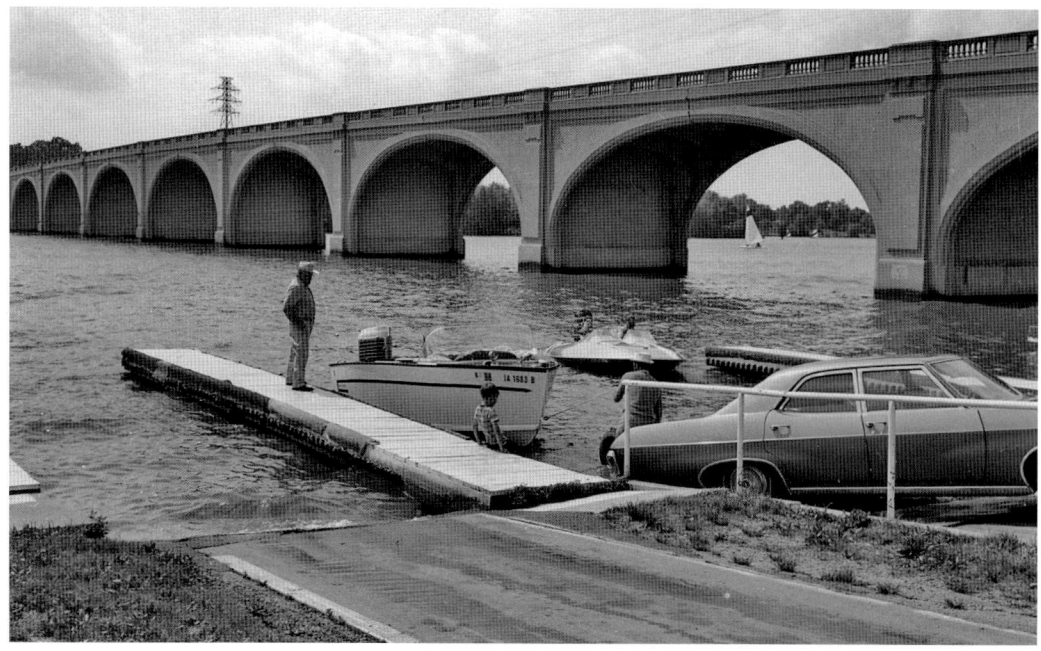

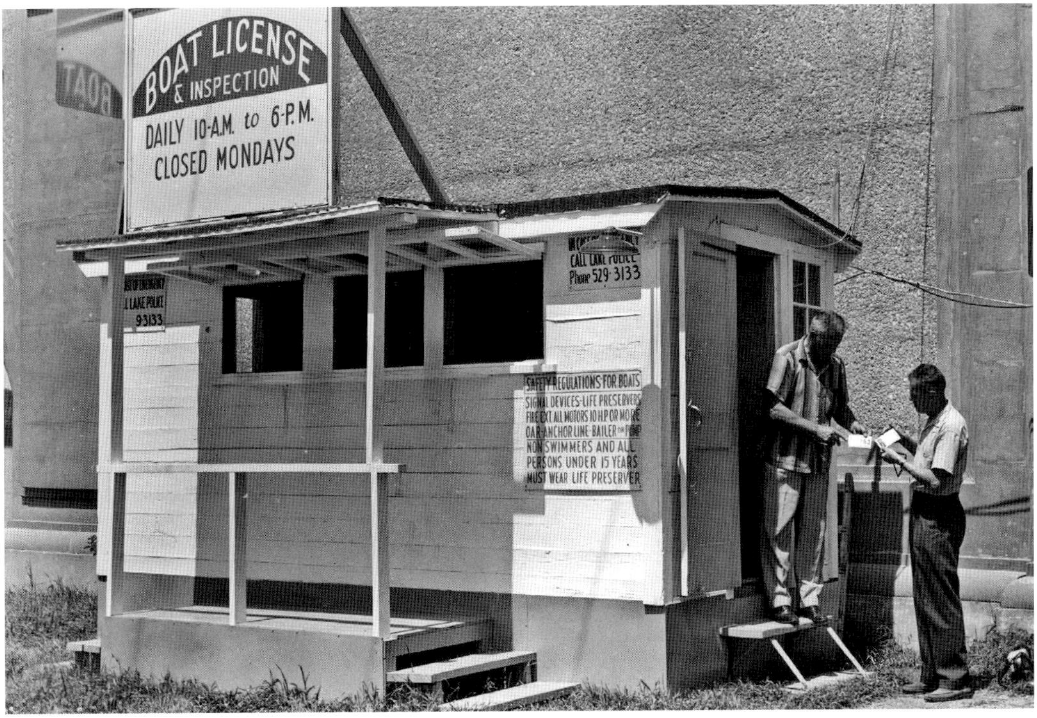

Of course, with recreation and public access comes law and order. A small boat license shack, built in the shadow of Lindsay Bridge, allowed boat owners to seasonally license their craft. The Lake Springfield Patrol served as the police presence on the lake, not only enforcing laws but providing assistance and rescue. [*Sangamon Valley Collection, Lincoln Library*]

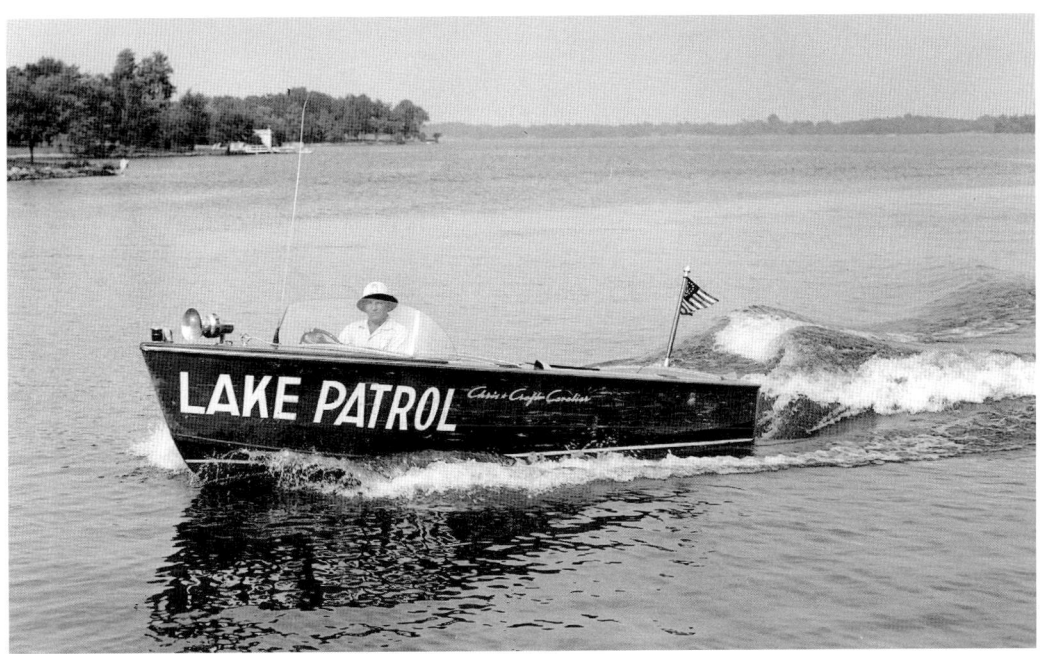

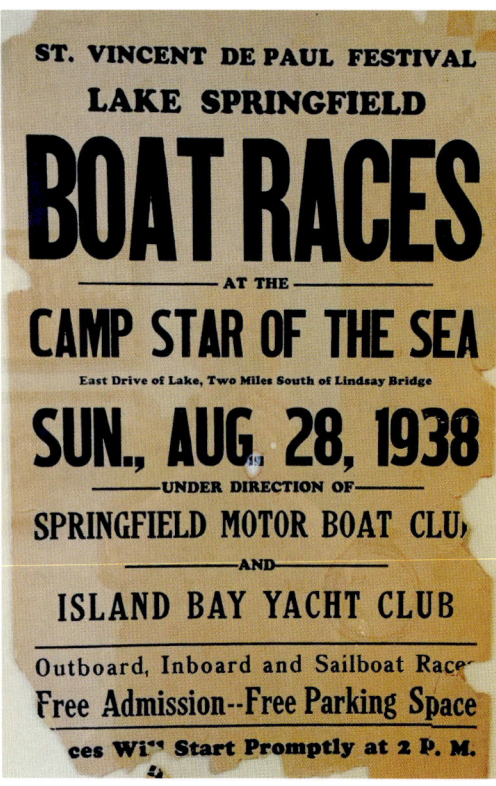

Motorboat racing on the lake was quite popular before World War II, and some races included dozens of boats that traversed the length of the lake several times. Such races are less popular today, and are generally confined to the western section of the lake. Below, Frank Mazrim's racing boat, *Jezebel*, is tagged and ready for one such race, *circa* 1953. [*Courtesy of Robert Mazrim*]

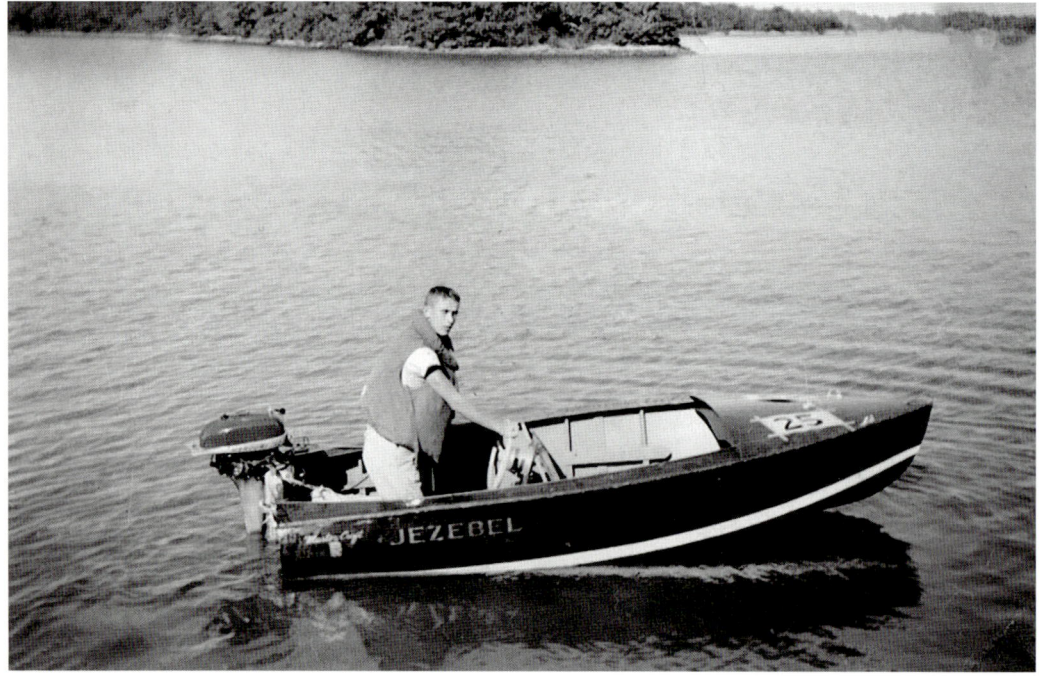

Things to Do

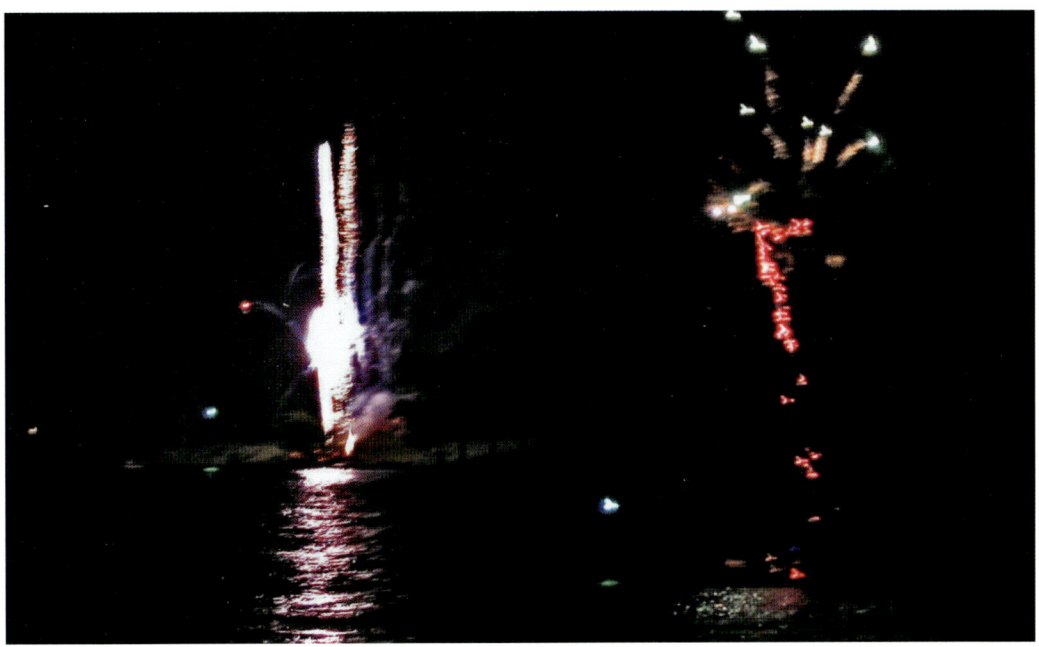

The Fourth of July has been one of the most popular holidays on the lake since the 1960s. Hosting not only parties and concerts, a number of boat clubs have sponsored fireworks displays. Meanwhile, residents of the lakeshore put on their own displays (above), causing the water to be lit with hundreds of colorful explosions during the evening. Below, a photo inadvertently captures the reaction of one celebrant during a bottle rocket misfire. [*Courtesy of Robert Mazrim*]

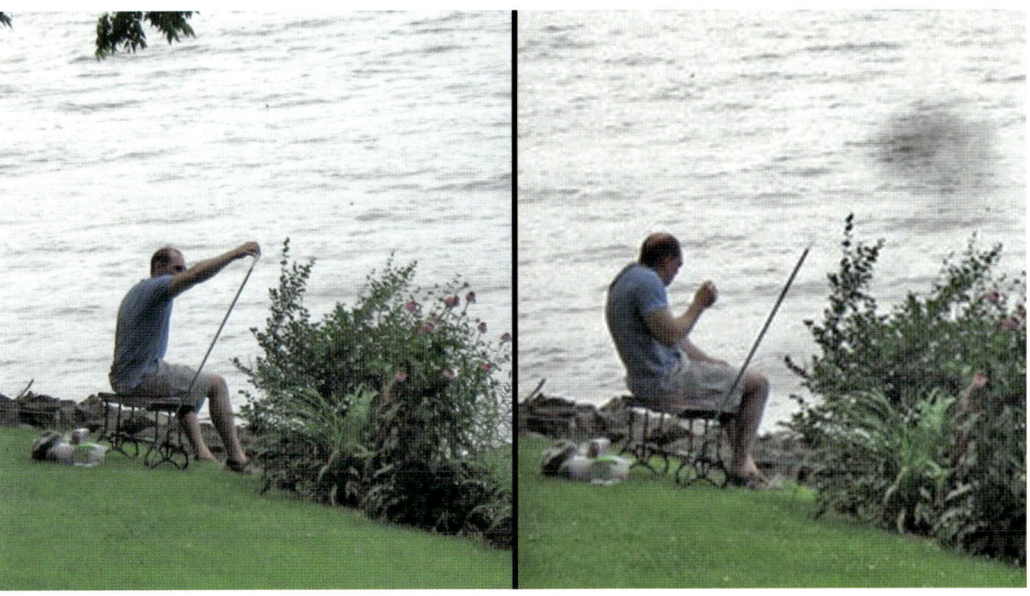

6

At the Clubs

Between 1935 and 1965, nearly two dozen private clubs were established along the shores of lake. Of course, most of these were intended to provide access to the water for boating, skiing, and swimming—often for people who did not own lake property. In some cases, the clubs were fraternal or occupational in theme: the Masonic Club, the Knights of Columbus Club, the Jesters Club, the Postal Club, the Press Club, the Amvets Club, or the TRN Club—which was established for members of the television, radio, and newspaper professions. The Prop Club was formed by employees of the Allis Chalmers plant in Springfield. Others, such as the Surf Club, the Anchor Boat Club, the Ski Club, and the Rod and Reel Club focused on water-themed activities.

Some clubs, such as the Springfield Motor Boat Club, were actually organized before the lake was completed and initially held their meetings and events elsewhere. The first clubhouses to be constructed on the lake included the Crab Apple Club, the Press Club, the Island Bay Yacht Club, and the "Cuckoo Club"—located on the extreme western end of the lake and known primarily for hosting the Robert Burns Club meetings during the early 1940s.

The first actual clubhouse on Lake Springfield was probably that of the Crab Apple Club. Located in Orchard Lane on the south shore, the clubhouse consisted of an 1870s farmhouse that was originally part of a farm that had been recently flooded by the lake. The club had only a few amenities, including a boat dock, some horseshoe pits, and a picnic area on a point overlooking the water. Aside from club gatherings, the old house was rented to other local organization for meetings and picnics during the 1930s and 1940s. The club closed in the mid-twentieth century, and the old farmhouse was eventually demolished. Old horseshoe pits (a common feature at most lake parks) could be seen in the backyard of the house for years after its abandonment.

The Victory Boat Club hosted impressive 150-boat, 50-mile races during the 1940s. Marcy's boat dock provided boat rentals, repairs, gasoline, and a snack bar open to the general public (requiring no membership), and also hosted early boat races. Island Bay Yacht Club, which in its expanded form (constructed in 1966) represents one of the largest clubhouses on the lake, has always focused on sailing. The club has hosted regular sailboat regattas since 1939.

At the Springfield Ski and Boat Club, members and the public alike watched trick skiers vault over a floating ramp, or slalom ski through a competitive course of buoys anchored in the water in front of the club. The Illinois State Water Ski Championships were held there as well. The Lake Shore Club featured numerous dances, dinner parties, and a Hawaiian luau in June of 1958. The grounds of that club, which still boast an Olympic-sized swimming pool, were sold to the Elks Club in 1979.

In 1937, the Girl Scouts of America created Camp Widjiwagan on 76 acres of wooded ground on the western end of the lake, overlooking the former Lick Creek Valley. Across this narrow portion of the lake is Camp Illinek, established by the Boy Scouts of America two years later. Camp Star of the Sea was founded by St. Joseph's Catholic Church in 1935. The camp later became the site of Villa Maria, an educational facility for diocesan seminarians. Camp Star of Sea was built on a portion of the historic Christopher Newcomer farm, and the Newcomer family cemetery can still be seen along East Lake Drive.

Near the Boy Scout camp on the western end of the lake is the Lick Creek Wildlife Preserve, a 34-acre natural area near the headwaters of the lake originally set aside to protect the water quality through erosion control. The area was first known as "The Arboretum." Groves of sugar maples and chinquapin oaks are found in the forests there, and outcrops of limestone form parts of the lakeshore in this part of the lake.

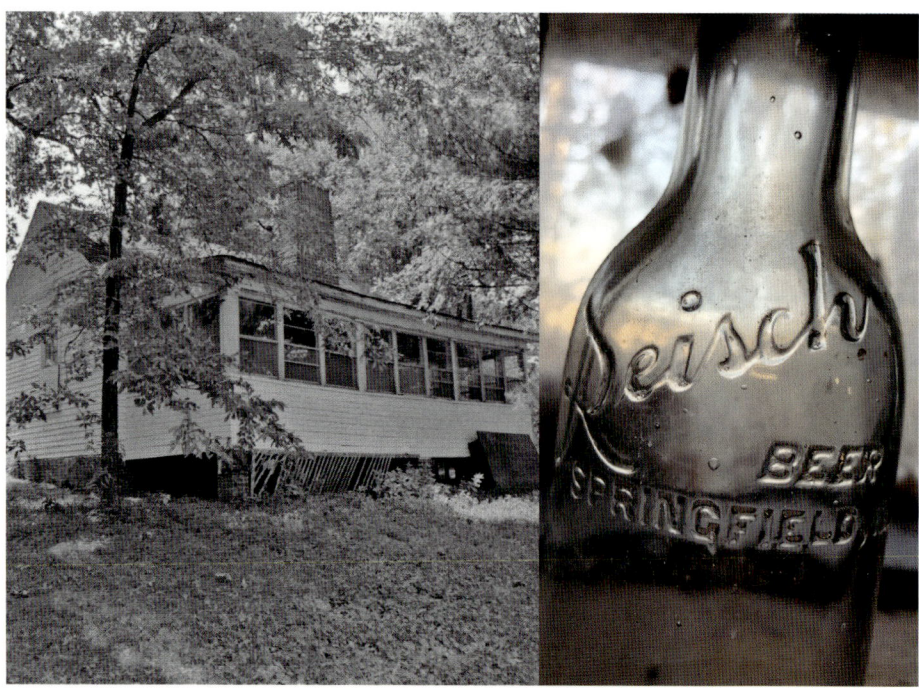

One of the earliest clubs to be established along the shore was the Crabapple Club, which was housed in a nineteenth-century farmhouse that once overlooked Sugar Creek. When it was demolished in the 1980s, the old well behind the house was found to be filled with artifacts from good times held at the club. Shown below is the original clubhouse affiliated with the Island Bay Yacht Club, the club principally focused on sailing at Lake Springfield. [*Above: Courtesy of Robert Mazrim; Below: Sangamon Valley Collection, Lincoln Library*]

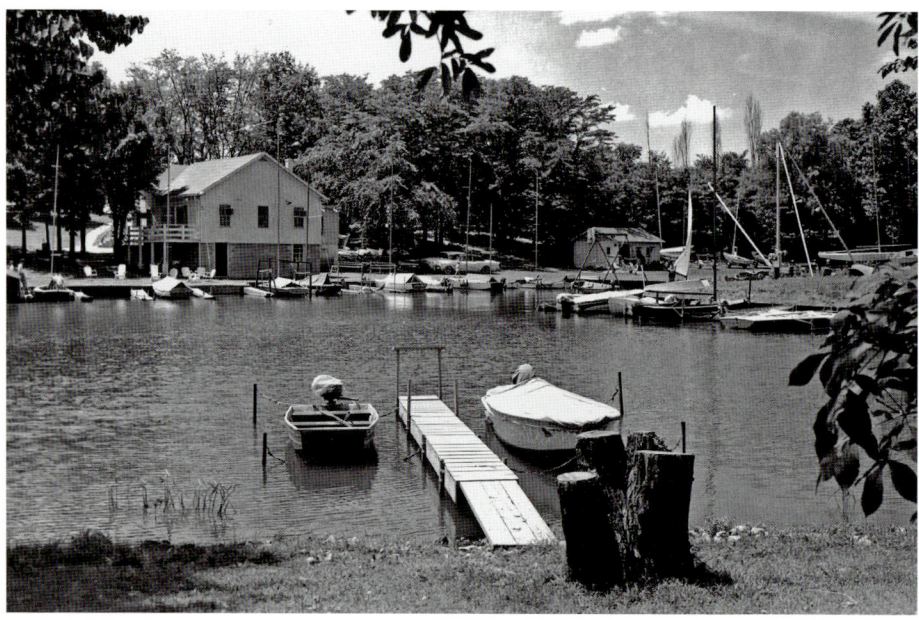

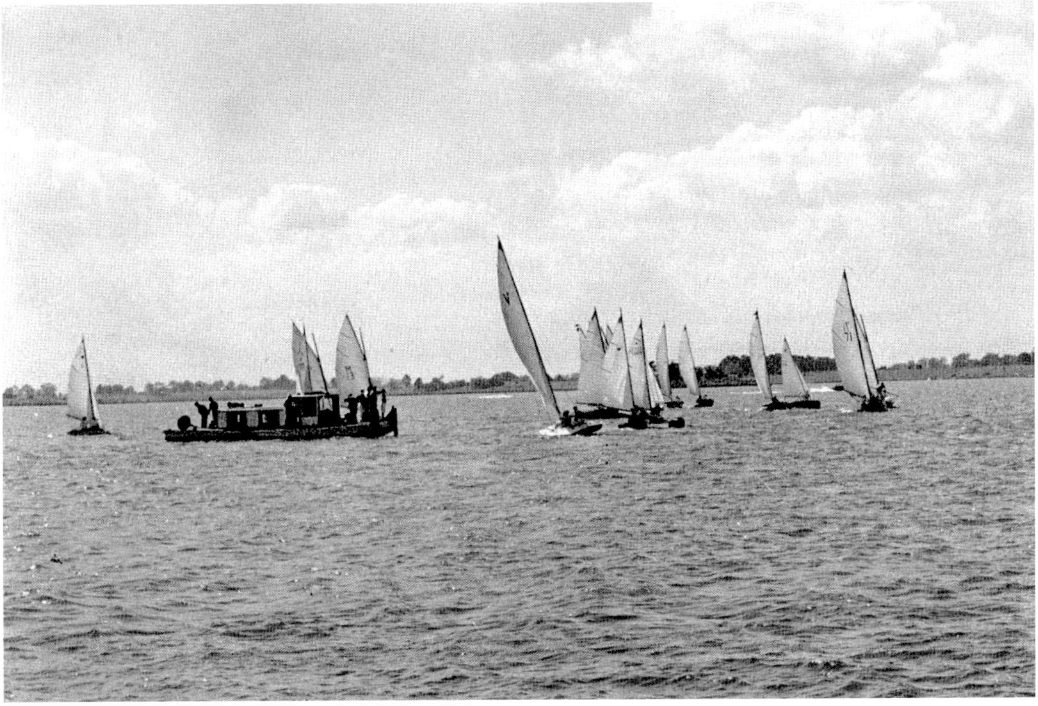

At the Clubs

Above is a photograph of one of many of the Island Bay Yacht Club regattas, held in the open water in what is the largest part of the reservoir. Below is the impressive replacement of the original Yacht Club clubhouse, constructed in 1966. [*Sangamon Valley Collection, Lincoln Library*]

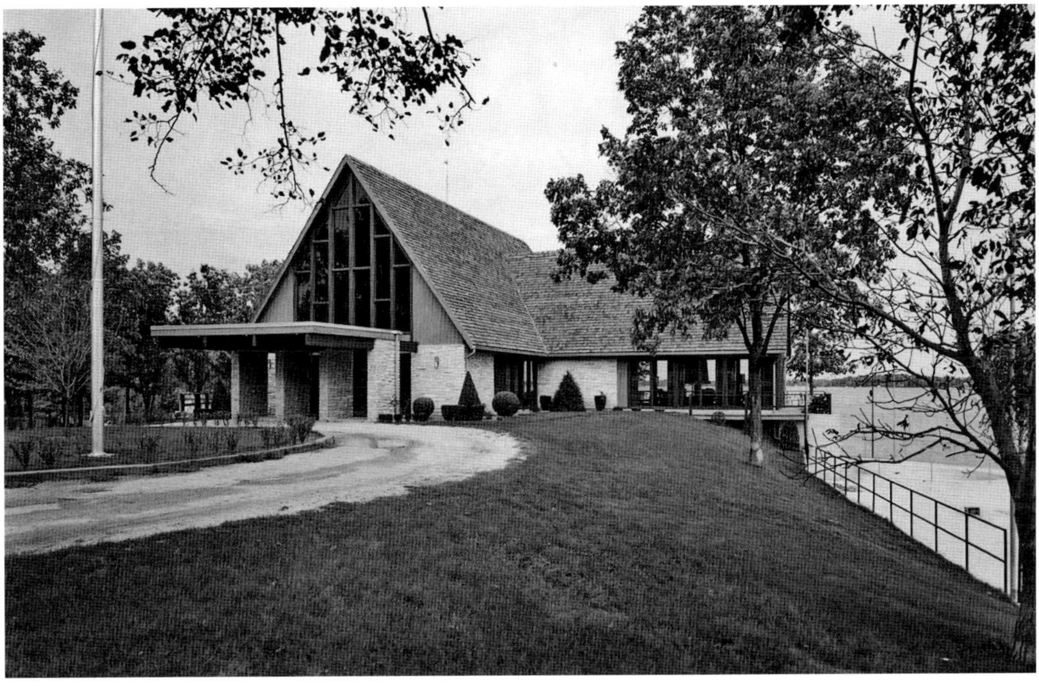

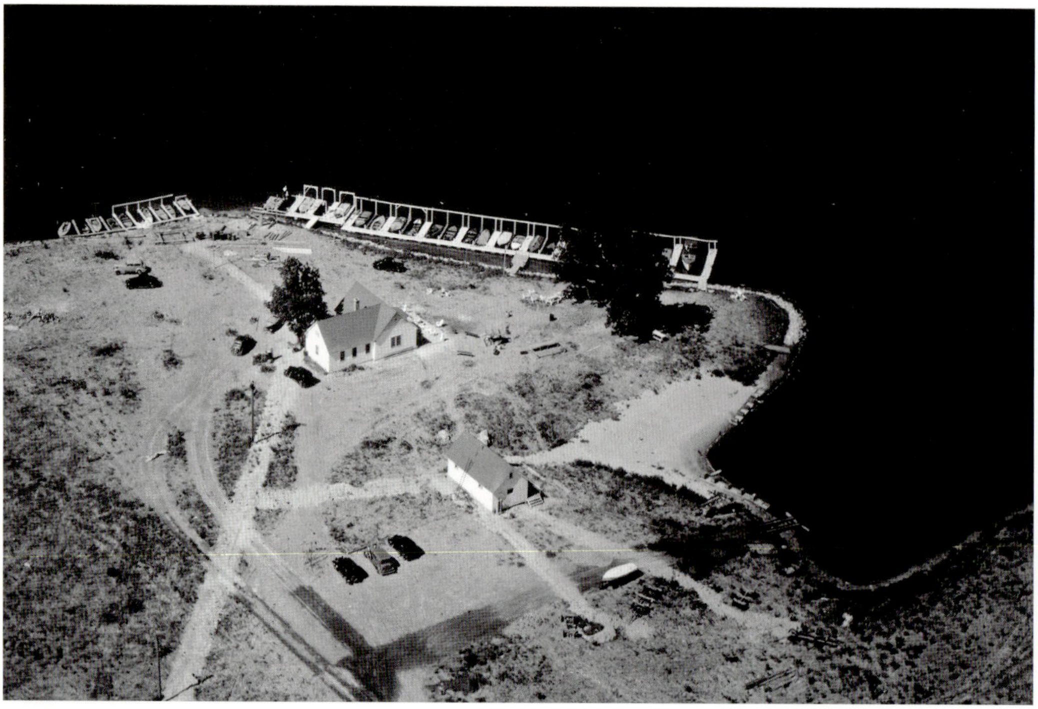

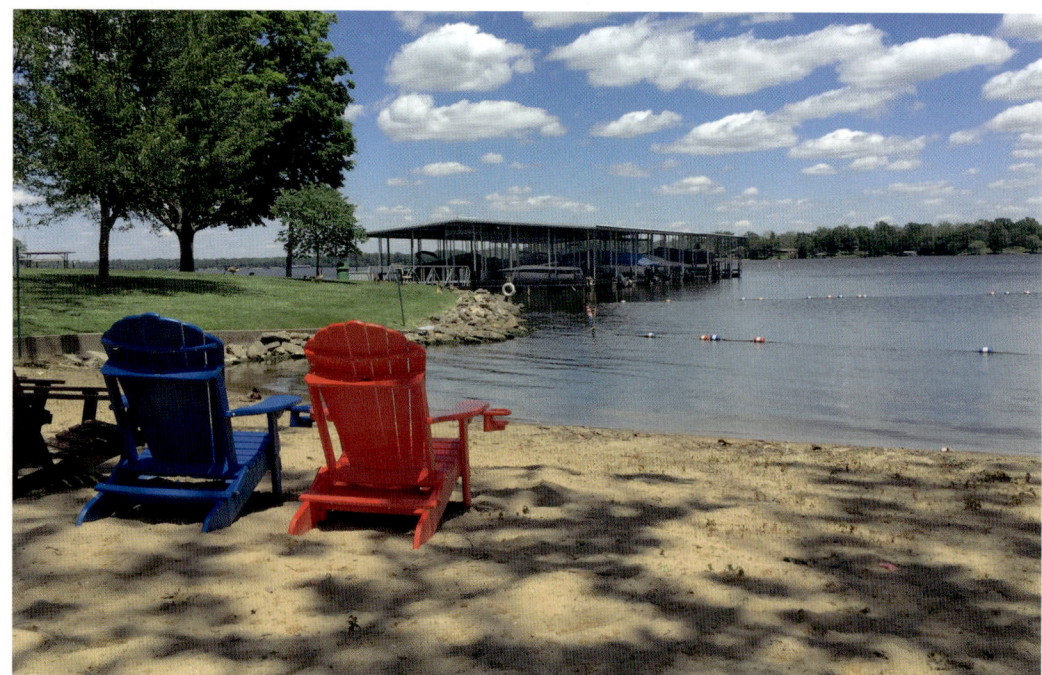

Aerial view of the recently constructed Springfield Motor Boat Club, another early lake club that is still popular today. The beach at the Motor Boat Club looks much as it did sixty years ago. [*Above: Sangamon Valley Collection, Lincoln Library; Below: Courtesy of Robert Mazrim*]

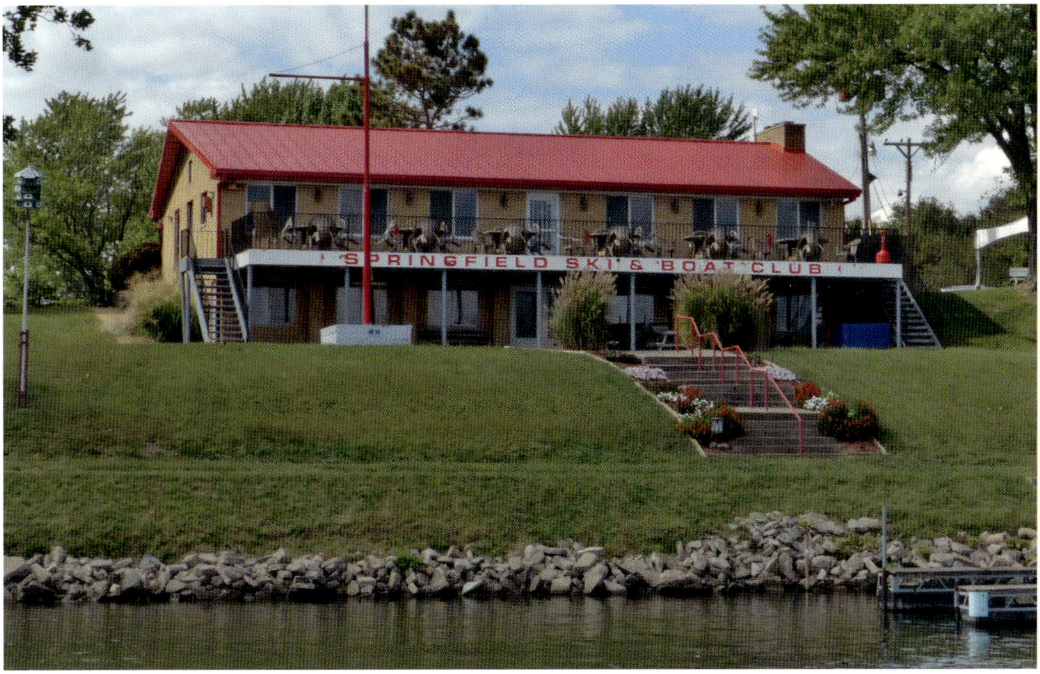

At the Clubs

The Springfield Ski and Boat Club is known for its water skiing competitions and exhibitions by trick skiers, such as those pictured below. The club was organized in 1957 by a small group of water enthusiasts and quickly grew in membership. [*Above: Courtesy of Robert Mazrim; Below: Sangamon Valley Collection, Lincoln Library*]

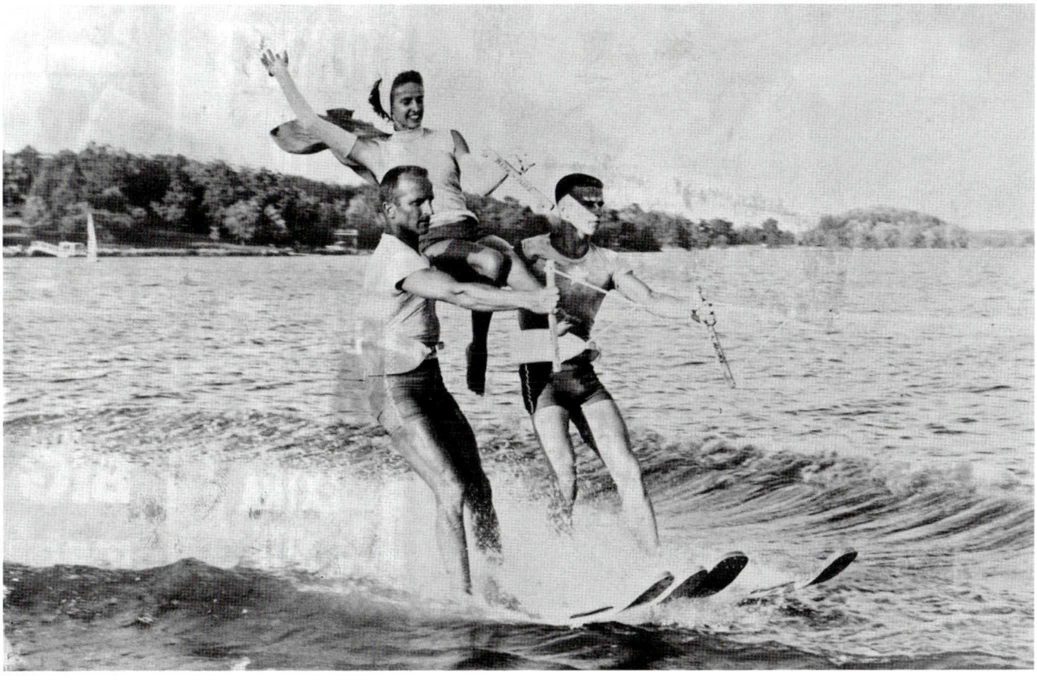

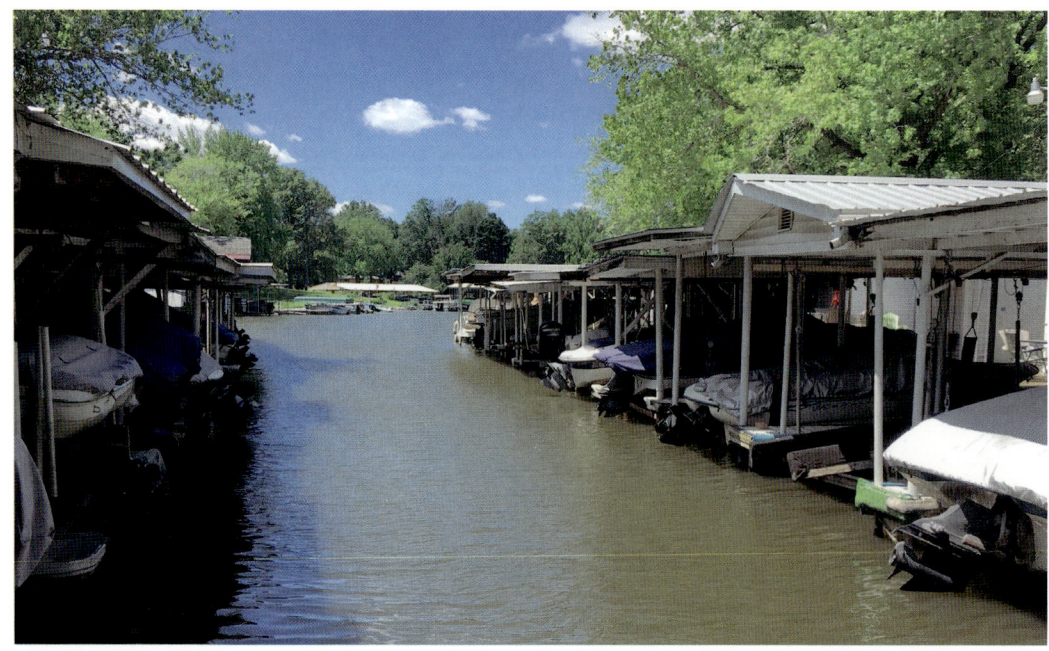

One important function of the boat clubs was to provide docking and storage for boats owned by those who did not live on the lake itself. At some clubs, the long winding rows of slips and small boat houses take on the appearance of tiny villages. Below is the interior of the Disabled American Veterans Club, replete with bar, fireplace, pool table, and knotty pine interior common during the 1950s. [*Above: Courtesy of Robert Mazrim; Below: Sangamon Valley Collection, Lincoln Library*]

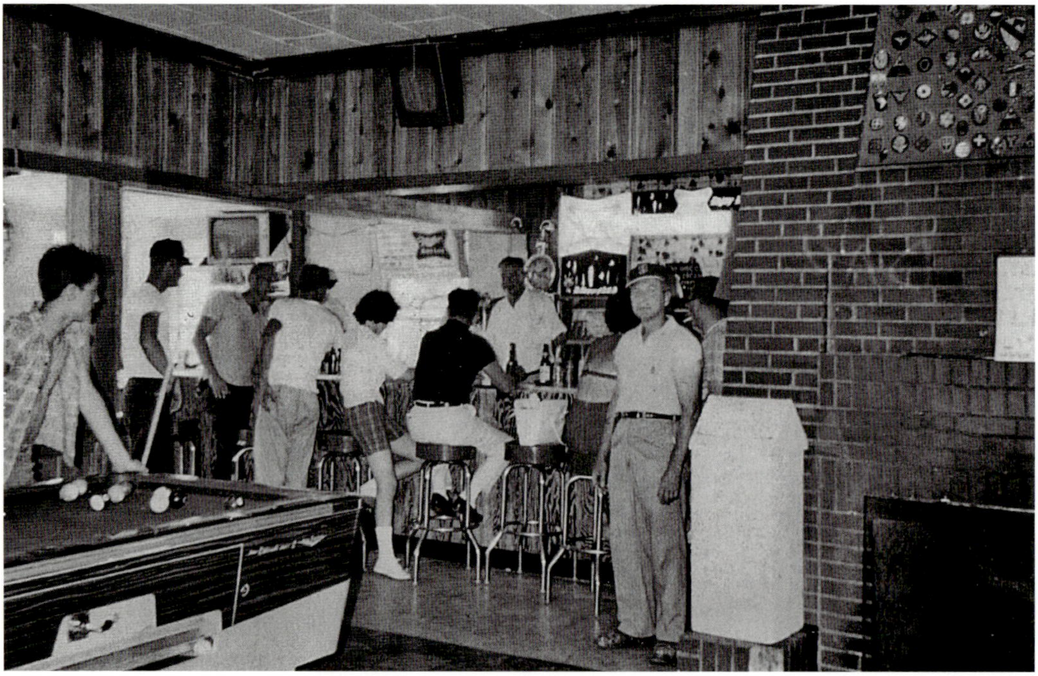

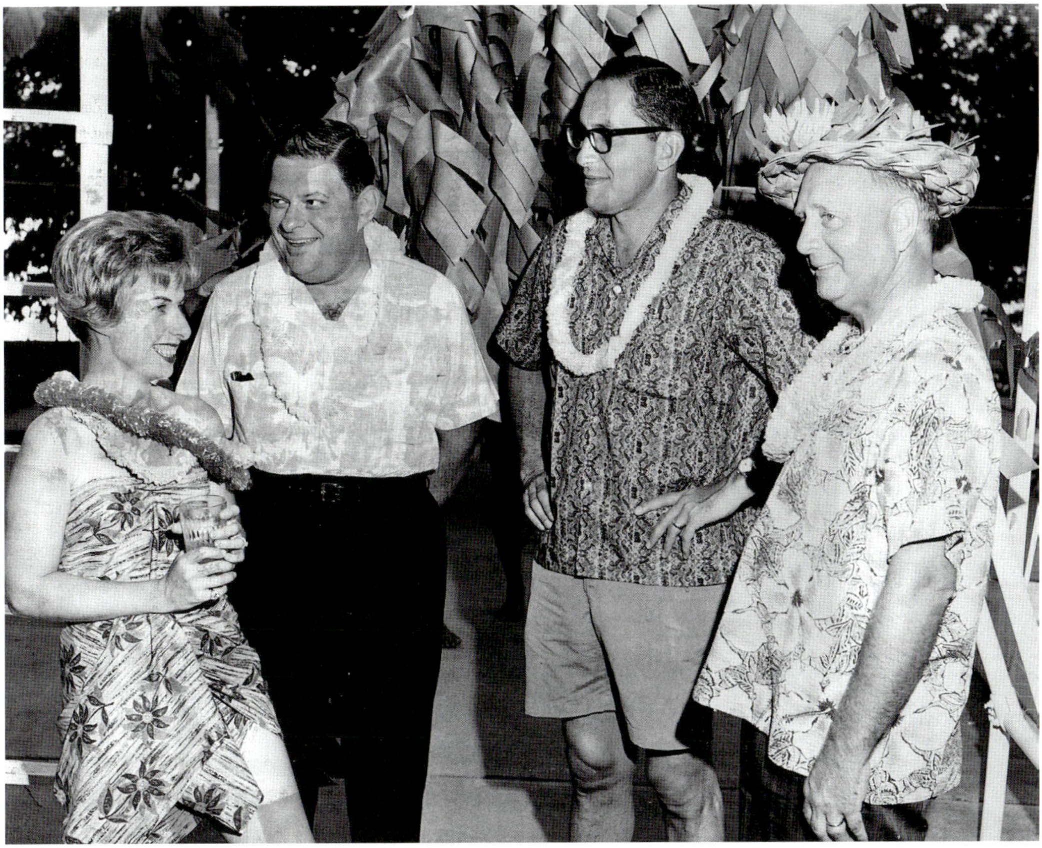

Members of the Lake Shore Club (now the Elks Club) enjoy "Luau Night" in 1964. A seahorse logo still adorns the side of the Ski and Boat Club. [*Above: Sangamon Valley Collection, Lincoln Library*; *Below: Courtesy of Robert Mazrim*]

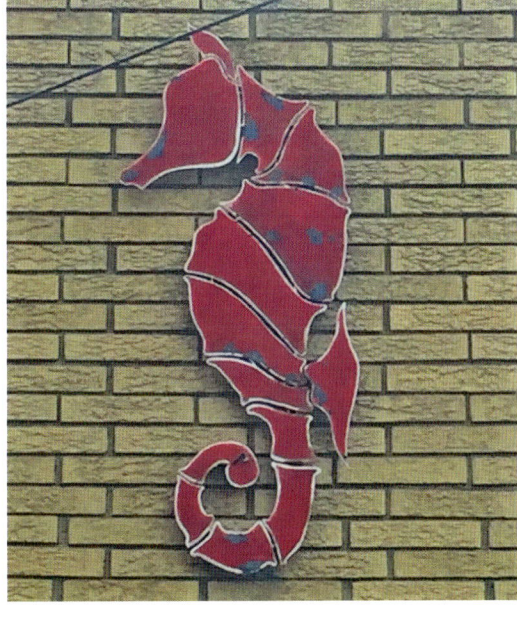

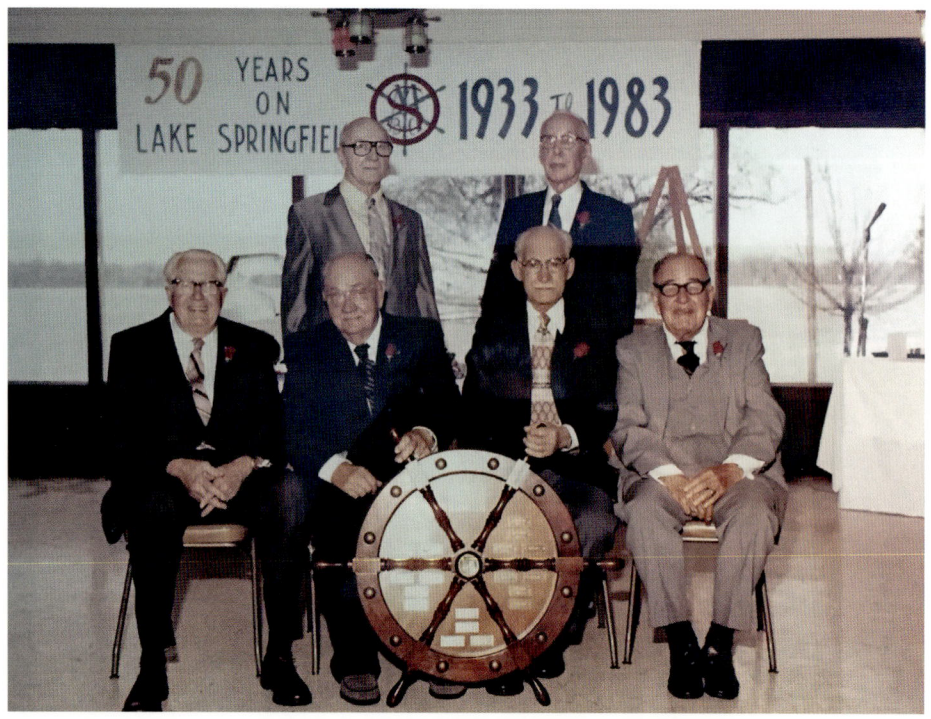

A photograph of Springfield Motor Boat Club board members taken on the fiftieth anniversary of the club in 1983. Below is an example of the kind of folk art found at some of the private lake clubs, including a mural at the Springfield Motor Boat Club, and a totem at the Jesters Club. [*Courtesy of Robert Mazrim*]

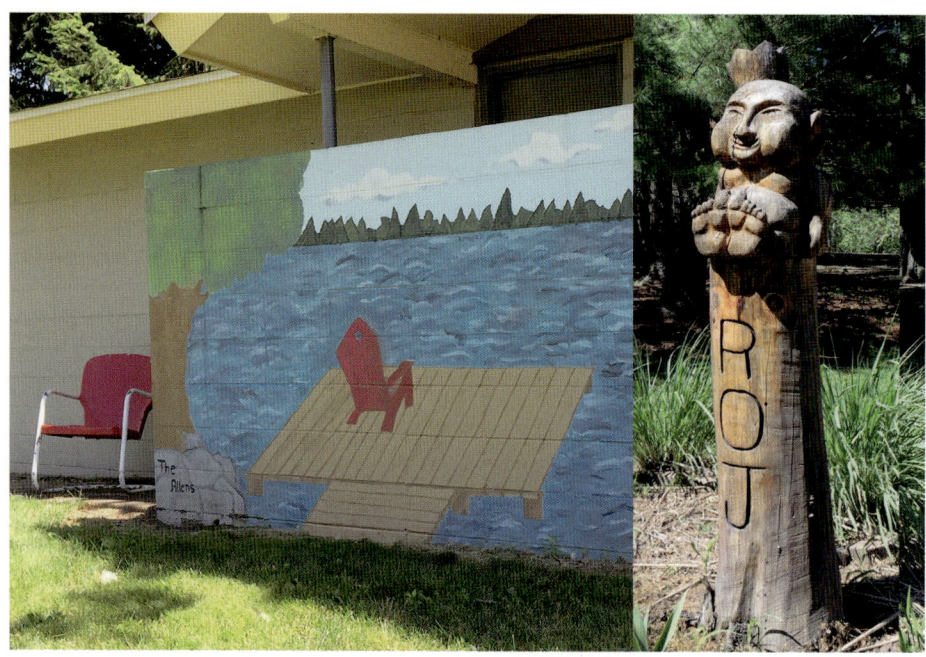

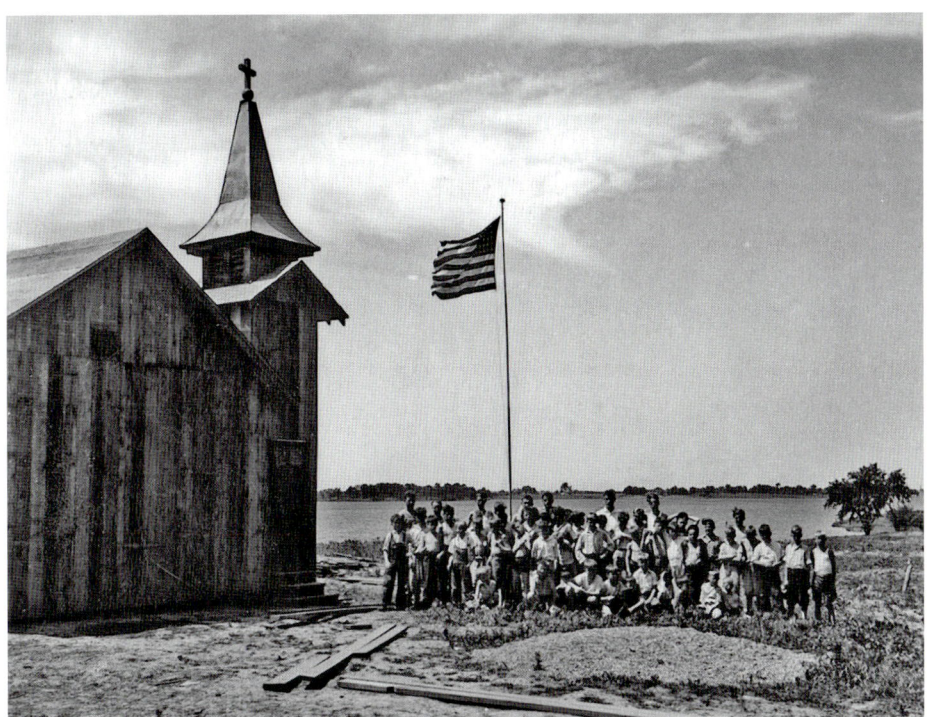

Camp Star of the Sea was founded by St. Joseph's Catholic Church in 1935. The camp remained opened until 1975. Today, only some of the lodging from the original camp still stands. [*Above: Sangamon Valley Collection, Lincoln Library; Below: Courtesy of Robert Mazrim*]

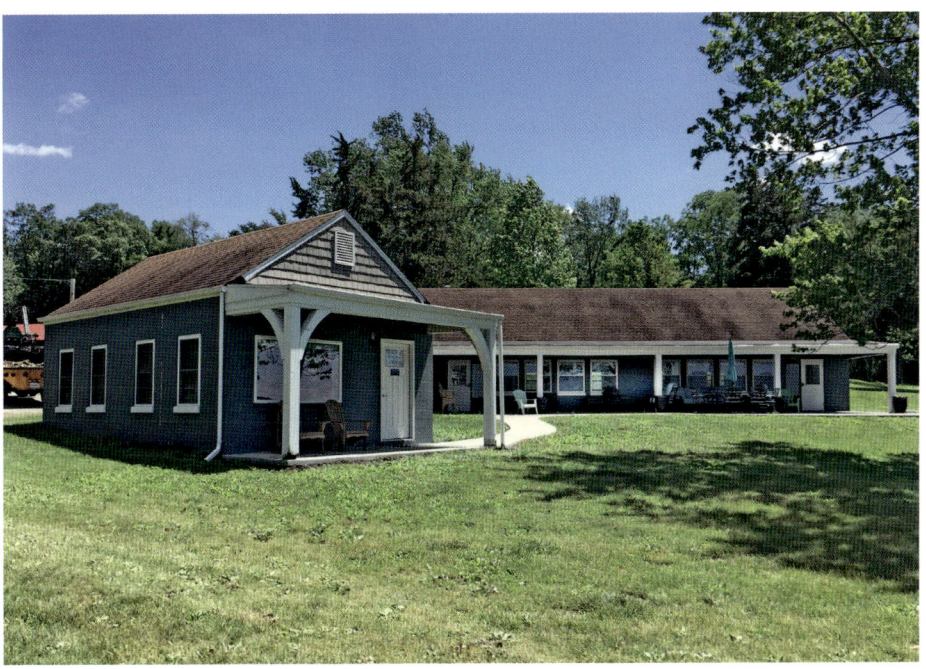

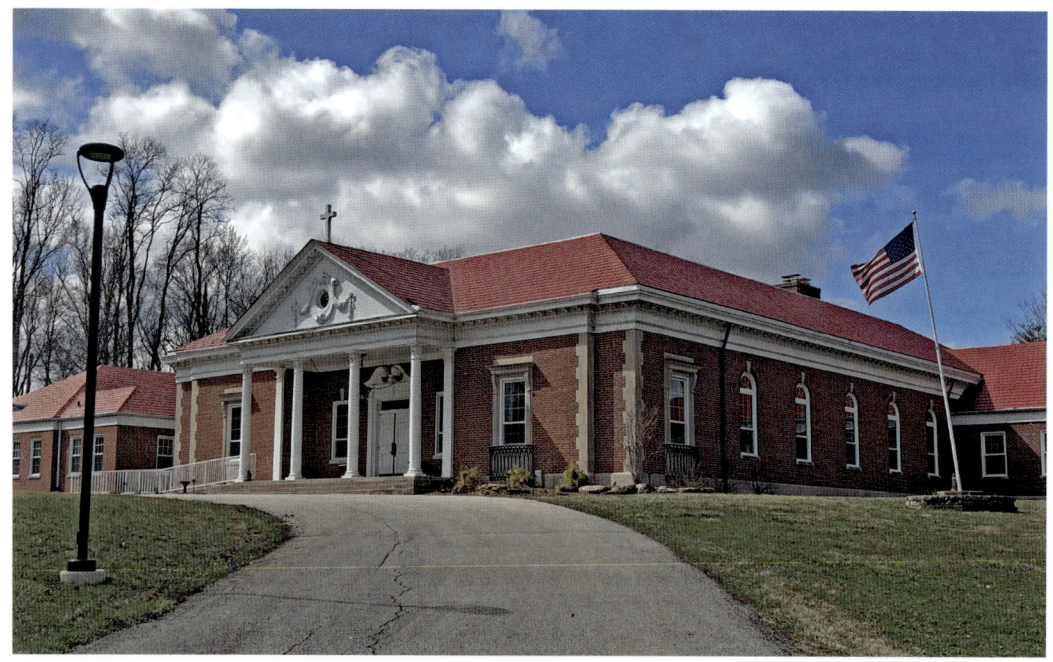

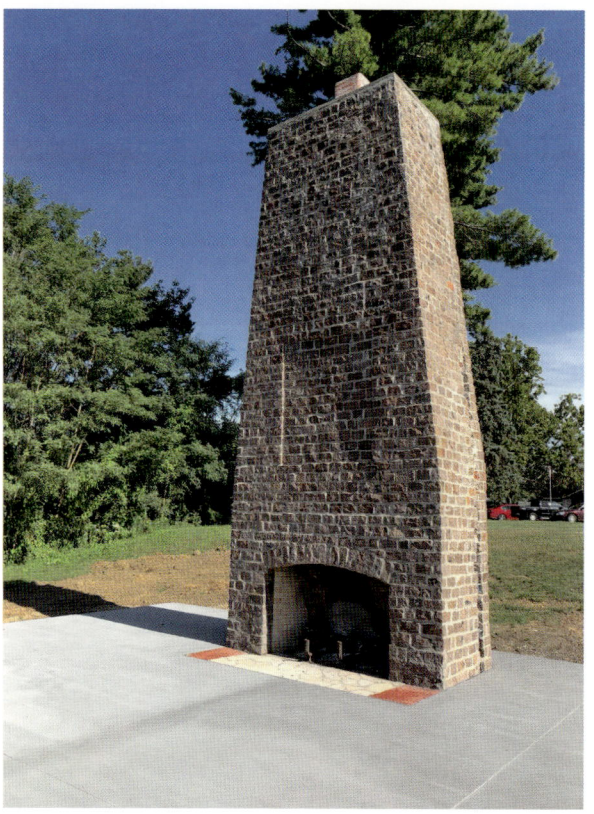

In 1955, The Latin School, a one-year preparatory school for future seminarians, became a neighbor of Camp Star of the Sea. Later, the school was known as the Diocesan Seminary of the Immaculate Conception. The site is now called the Villa Maria Retreat Center. Recently the lodge of the old camp was demolished, but the impressive chimney and fireplace were left standing on the grounds. [*Courtesy of Robert Mazrim*]

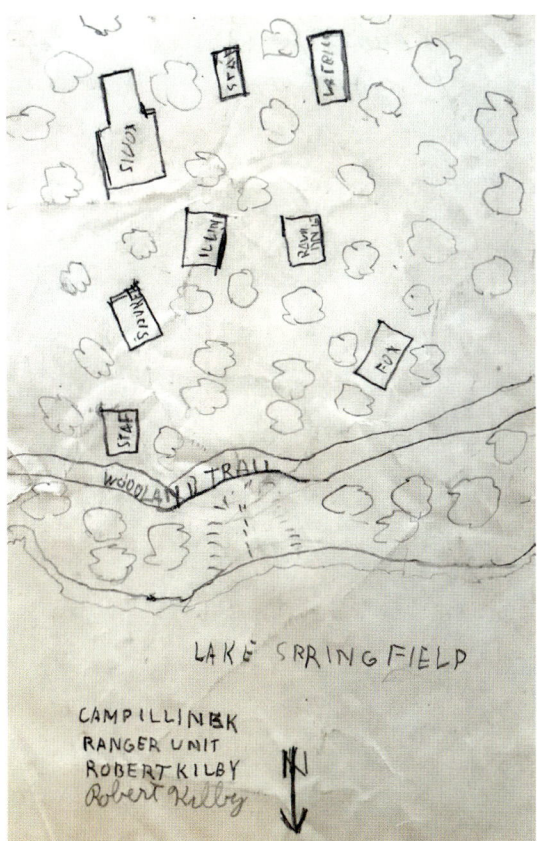

The Boy Scouts established Camp Illinek on the west end of the lake in 1939. This map of the cabins was drawn by one the scouts in the 1940s. Below, scout Bill Fanning is treated to a birthday swim off the docks of the camp in 1942. [*Courtesy of Robert Mazrim*]

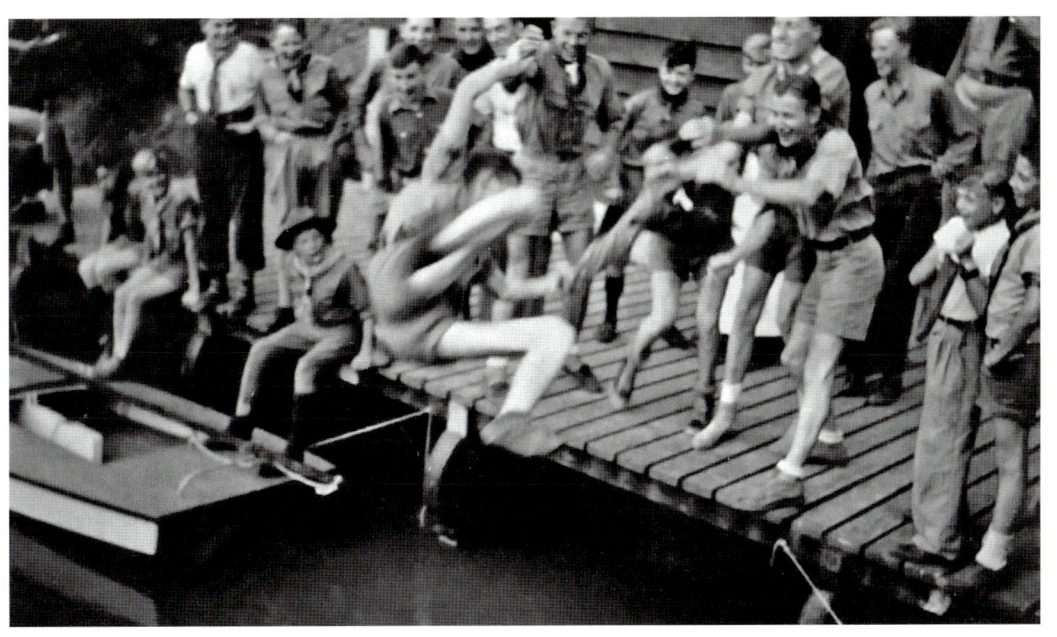

7

The Drought of 1952-55

Beginning in the fall of 1952, a severe drought began to cause the water level in the lake to fall. By mid-1953, water levels were alarmingly low, and by the summer of 1954, the city was facing a potential crisis. Beaches were closed due to a lack of water, and in places, the old Sugar Creek channel could be glimpsed for the first time in over ten years—surrounded by newly exposed mud that had been the lake bottom for a decade. Abandoned roadbeds were visible again, as were the remains of fencing, barn foundations, and tree stumps cut in the early 1930s.

In September of 1954, low water near the Route 66/South Sixth Street Bridge exposed a safe, a cash register, three guns, and costume jewelry, lying in the exposed mud. Lake police hoisted the loot from the top of the bridge, and assumed the items had been dumped into the water several months earlier from a "bandit truck." The next month, falling lake levels exposed a 22-foot steel lifeboat used by the Sea Scouts (at the Naval Training Station) for training on the water during World War II. In 1945, a storm tore it from its moorings and the boat disappeared. During the drought, a resident of Orchard Lane saw the bow of the boat sticking out of the mud in front of his home.

As water levels fell, Utilities Commissioner John Hunter considered the idea of seeding the clouds above the region with silver iodide to create artificial rain. At the same time, however, few water usage restrictions were enacted in the city itself. Water levels had fallen to more than 12 feet below normal pool. Pumps were installed to syphon water from the nearby South Fork Creek (which drained into the Sangamon River) in early 1955. That summer the drought lifted, and with the help of the new pumps, the lake levels began to rise again. The lake has not experienced such low levels since, but occasional low water still reveals traces of the Sugar Creek Valley's past.

The Drought of 1952-55

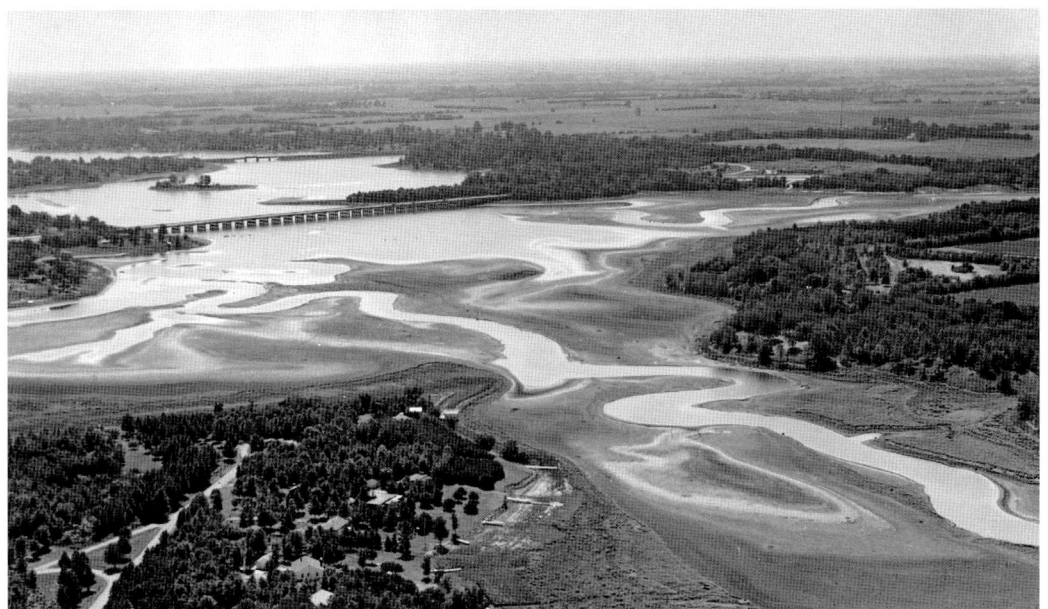

Between 1952 and 1955, an extreme drought dropped lake levels to an alarming degree. This aerial view, looking east, shows the old channel of Lick Creek (near where it merges with Sugar Creek) surrounded by a dry lake bottom. Below, a gradually receding shoreline resulted in boats beached many yards from their docks. [*Sangamon Valley Collection, Lincoln Library*]

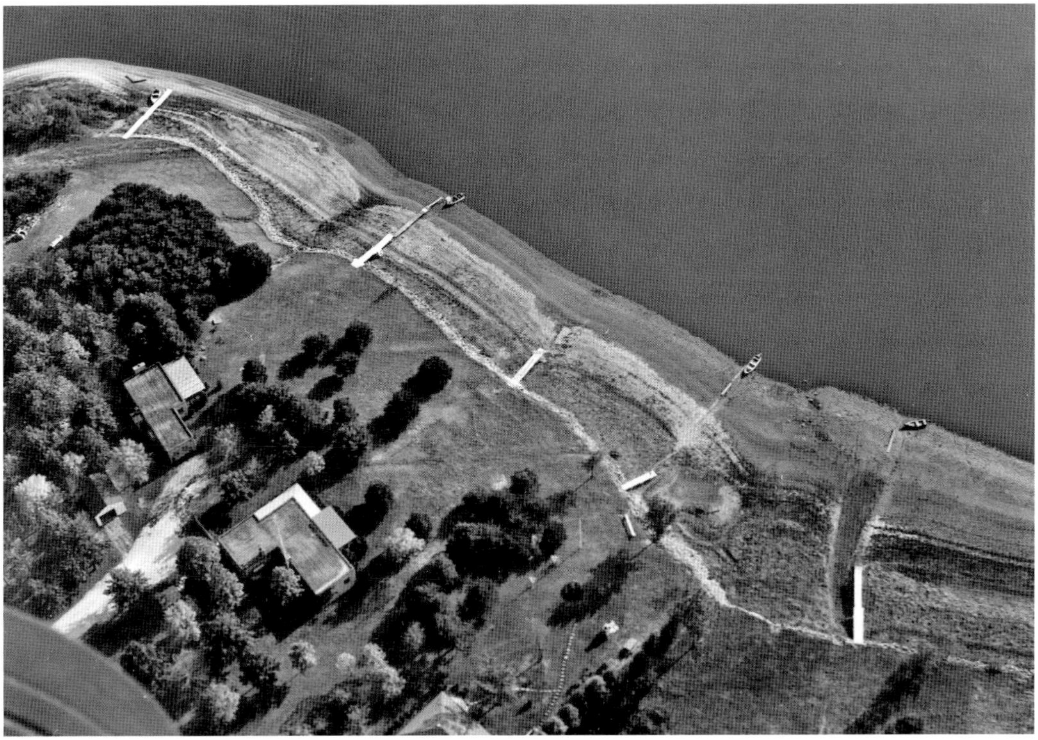

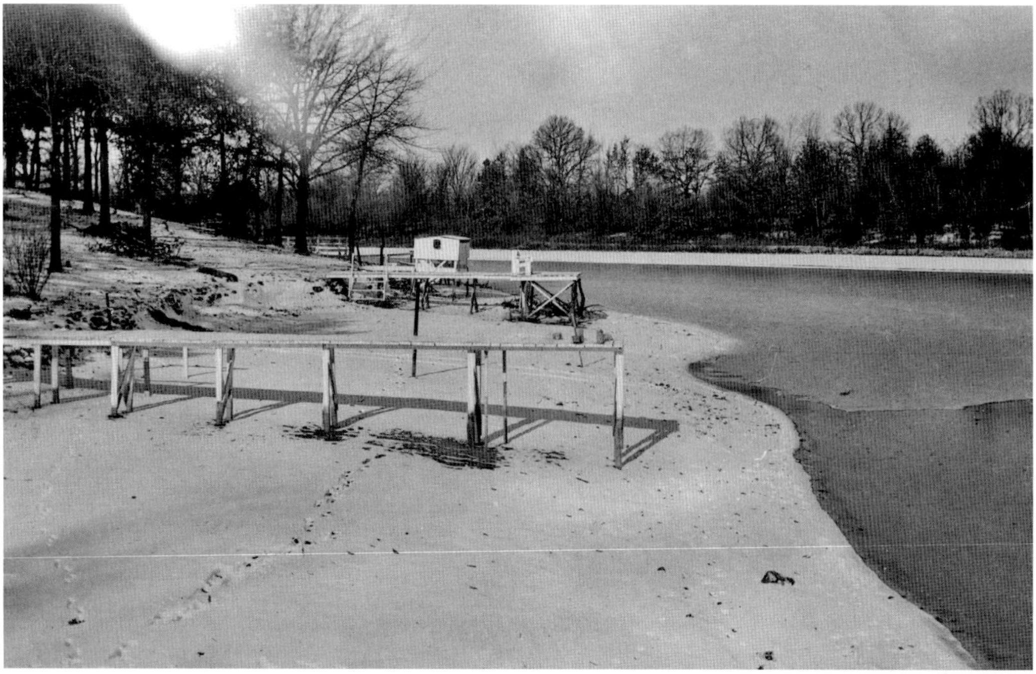

Lake Springfield in Illinois

These photographs were taken during the drought, in December 1953, along the shore of East Hazel Dell Lane. Above, docks stand completely out of the water. Below, an old floodplain terrace is exposed by low water. During occasional times of drought, that terrace has produced both early nineteenth century and prehistoric archaeological remains. [*Courtesy of Robert Mazrim*]

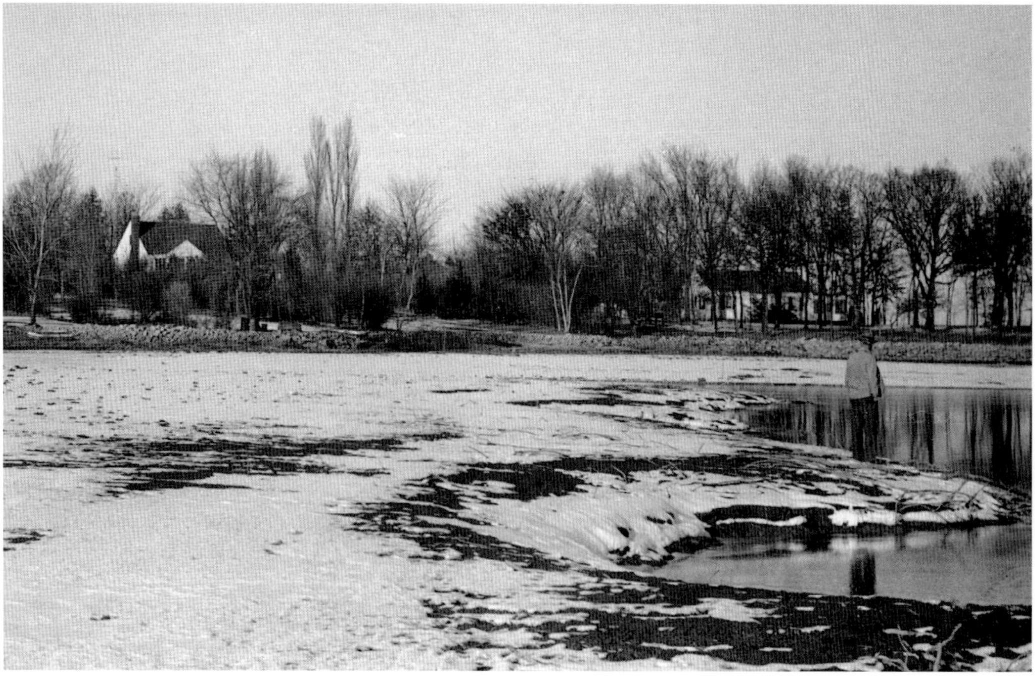

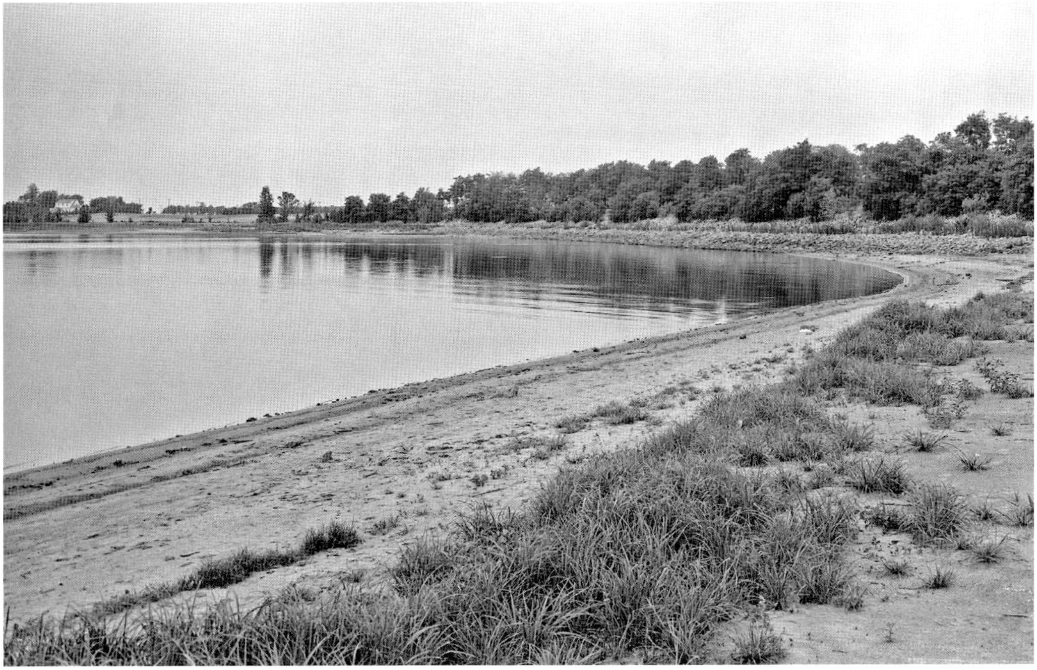

The Drought of 1952-55

The 1950s drought lasted long enough that residents of the lake became accustomed to the low water and extended beaches, some even planting melons in the sandy soil. Meanwhile, low water exposed a cache of "loot" dropped off of the Highway 66 bridge. Here, police hoist a safe and a cash register from the exposed mud. [*Sangamon Valley Collection, Lincoln Library*]

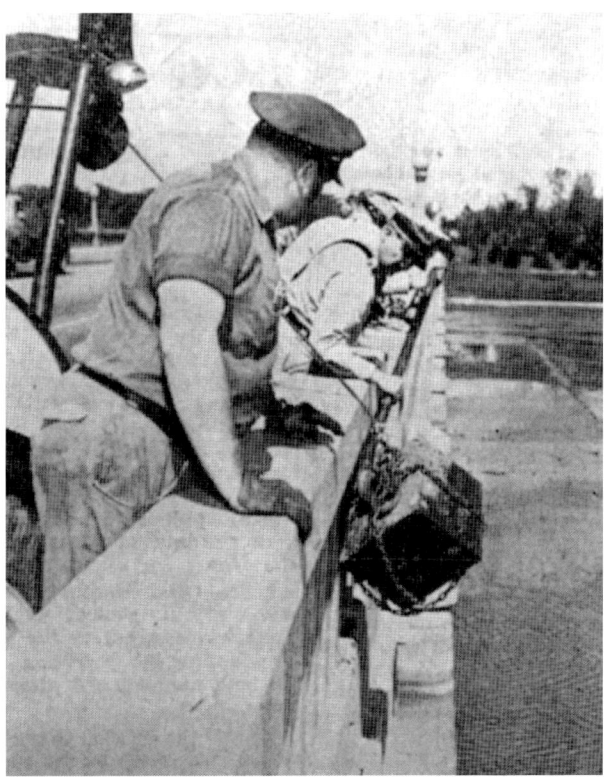

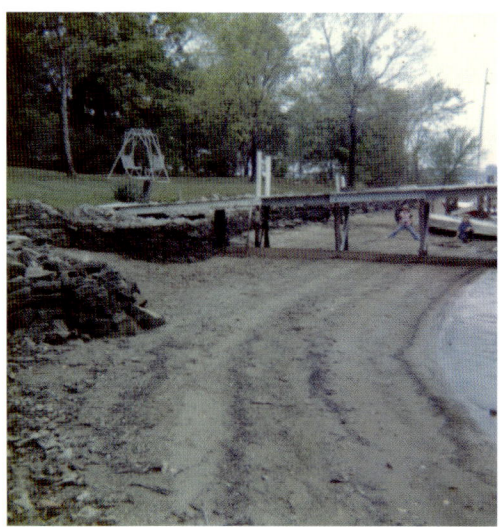 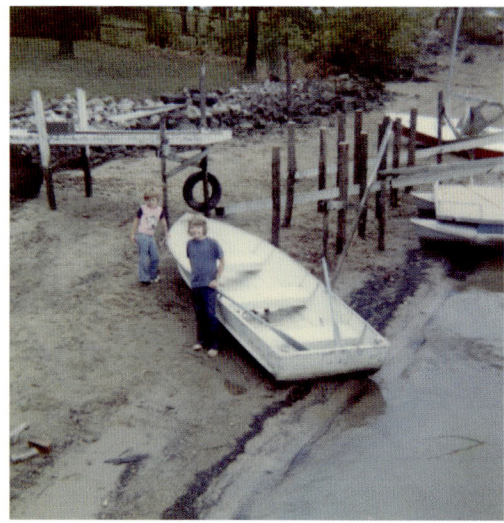

Although never as bad as the historic 1950s drought, low water occasionally returns to Lake Springfield. Here, during the mid-1970s, low water exposed sandy beaches in Orchard Lane. The owners of this property took advantage of that low water to build new docks on what was temporarily dry ground. Below, low water exposed a former hilltop that contained the remains of a nineteenth-century farmstead. Plainly visible was a concrete well cover, surrounded by domestic debris such as this bottle of "liver pills" that had spent forty years underwater. [*Courtesy of Robert Mazrim*]

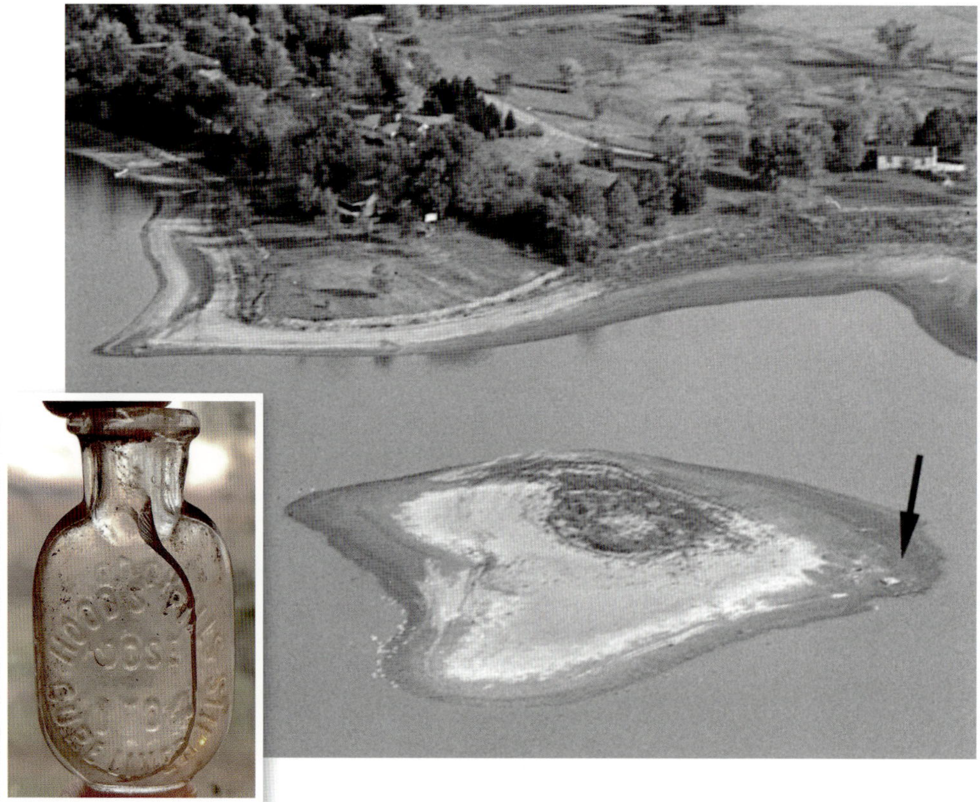

The Drought of 1952-55

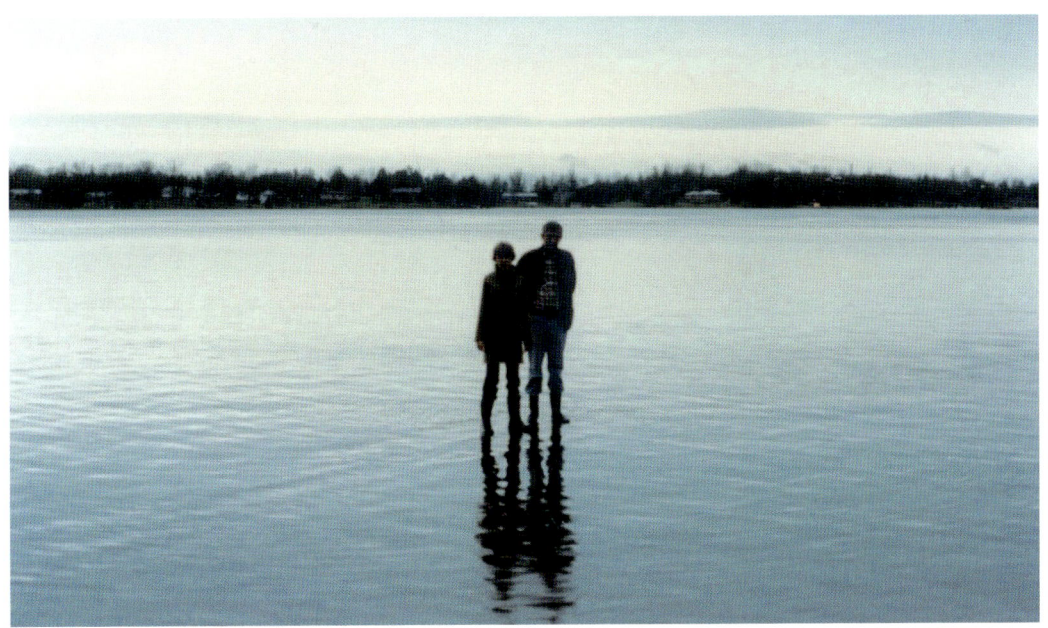

The lake has flooded at times as well. Here, in the early 1980s, residents of Orchard Lane appear to be walking on water, but in fact are standing at the end of a long dock submerged by floodwaters. Today, careful control of flood gates prevents serious flooding on the lake. [*Courtesy of Robert Mazrim*]

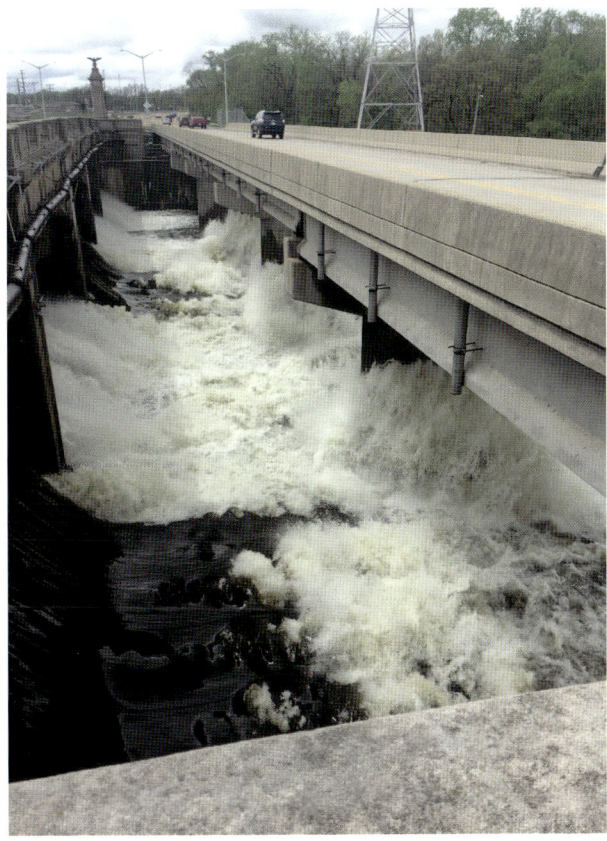

8

Life on the Lake

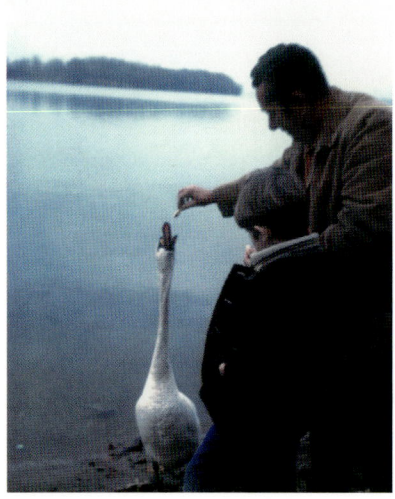

On Christmas Day 1972, this white swan was hand-fed on the shoreline of Orchard Lane. Children swimming off homemade docks is a less common sight than it once was. The snapshot below was taken around 1975. [*Courtesy of Robert Mazrim*]

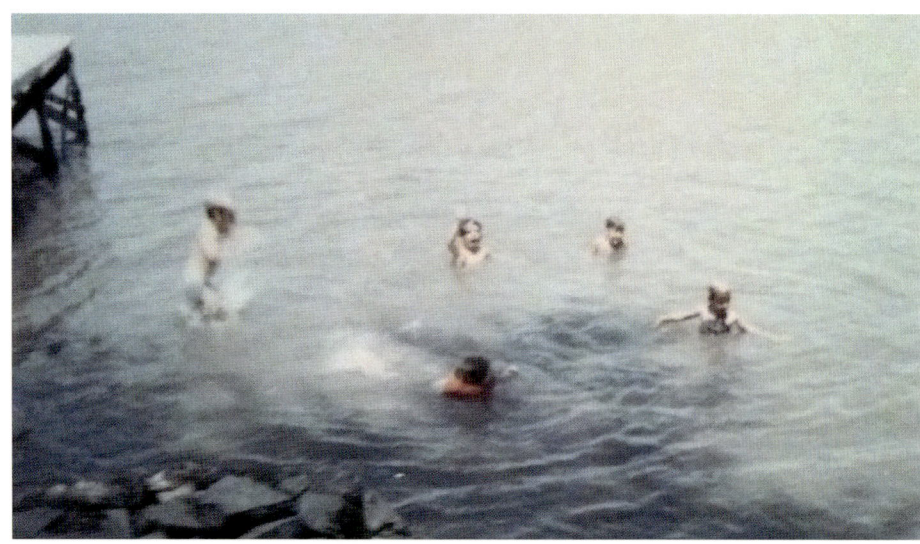

The new lake caused much excitement in the Springfield community in the 1930s. Downtown department store Myers Brothers advertised a series of window displays, each dressed in Lake Springfield themes. Below is an early advertisement for boat tours of the lake. For those who did not own a boat themselves, the *Miss Sasco* or the *Marianne* could introduce them to the waters of the new lake. [Sangamon Valley Collection, Lincoln Library]

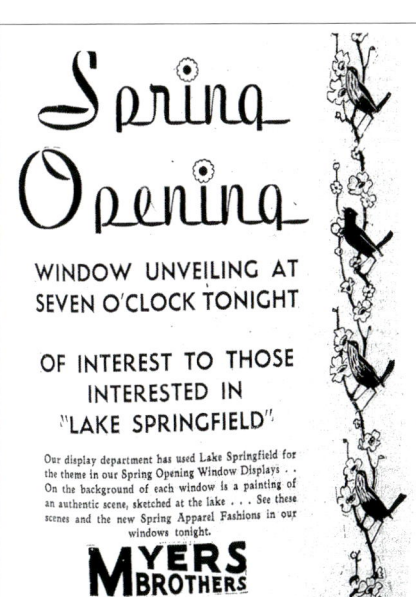

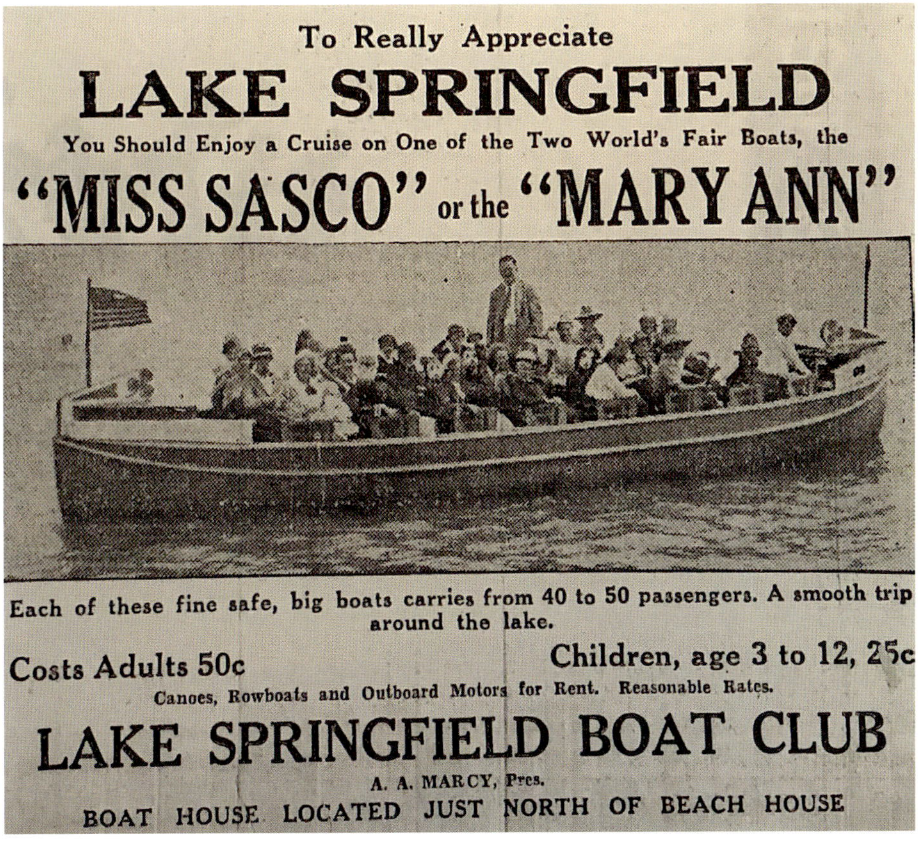

To Really Appreciate
LAKE SPRINGFIELD
You Should Enjoy a Cruise on One of the Two World's Fair Boats, the
"MISS SASCO" or the "MARY ANN"

Each of these fine safe, big boats carries from 40 to 50 passengers. A smooth trip around the lake.

Costs Adults 50c   Children, age 3 to 12, 25c

Canoes, Rowboats and Outboard Motors for Rent. Reasonable Rates.

LAKE SPRINGFIELD BOAT CLUB
A. A. MARCY, Pres.
BOAT HOUSE LOCATED JUST NORTH OF BEACH HOUSE

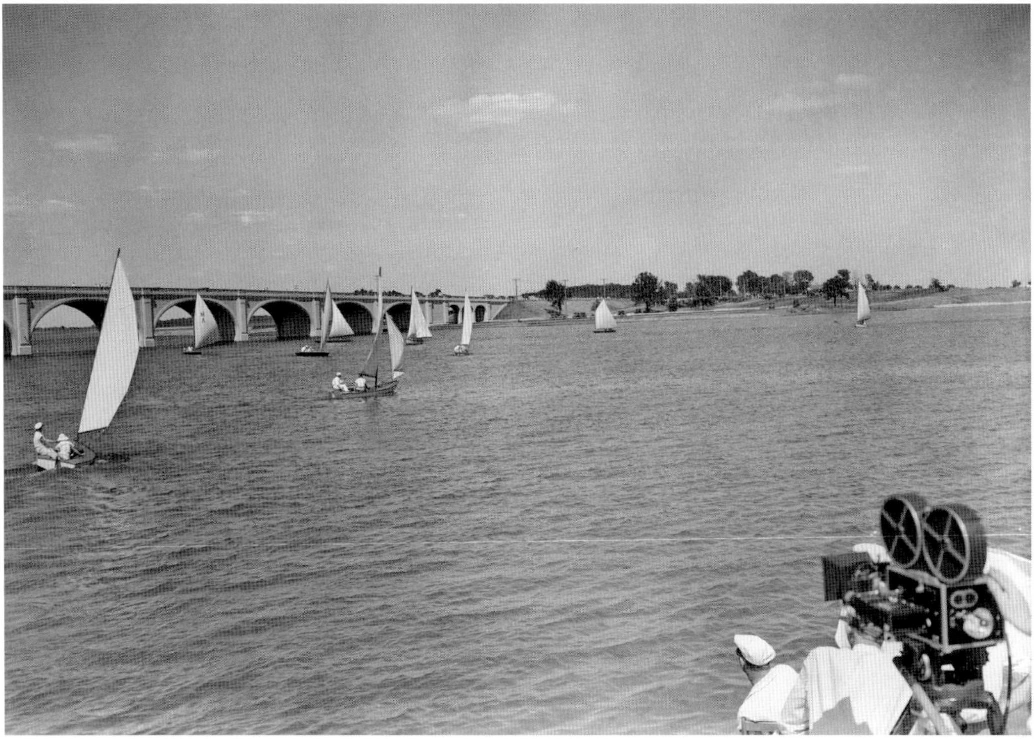

Lake Springfield in Illinois

Above, a film crew shoots a sailboat race near Lindsay Bridge. Below, a regatta at Island Bay Yacht Club. [*Sangamon Valley Collection, Lincoln Library*]

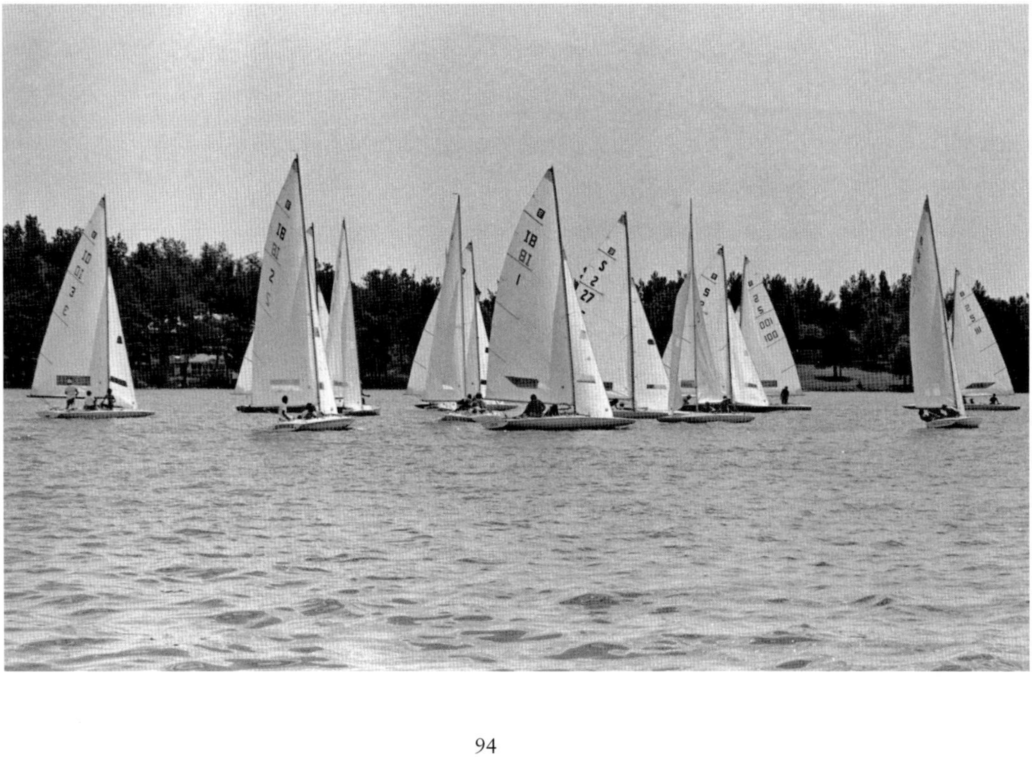

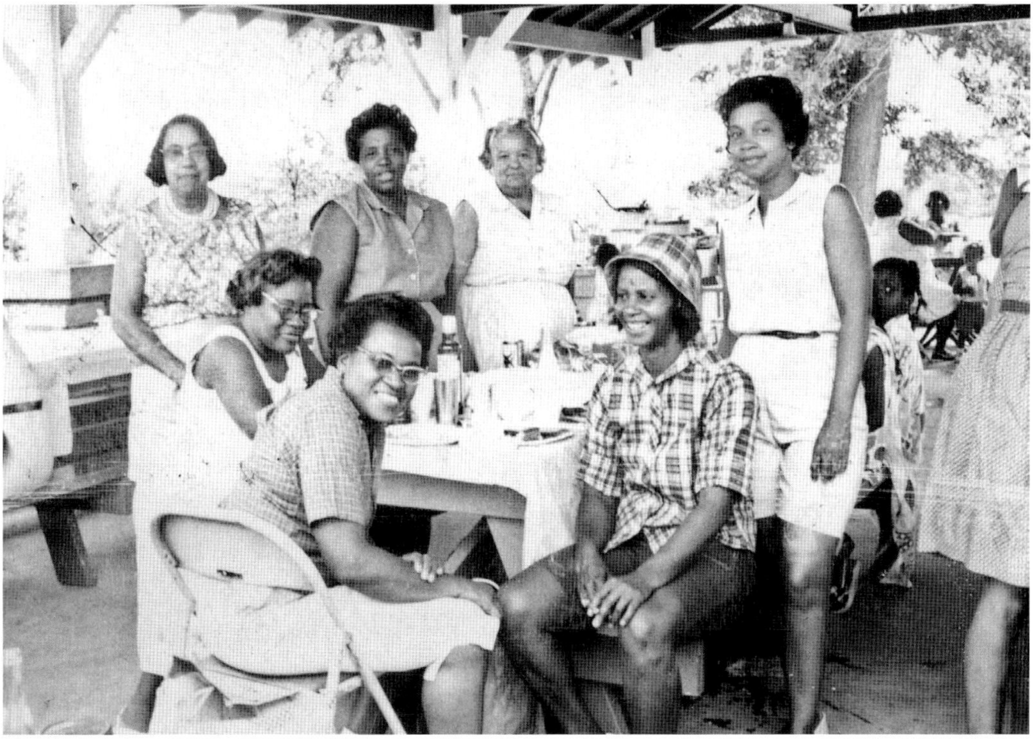

Life on the Lake

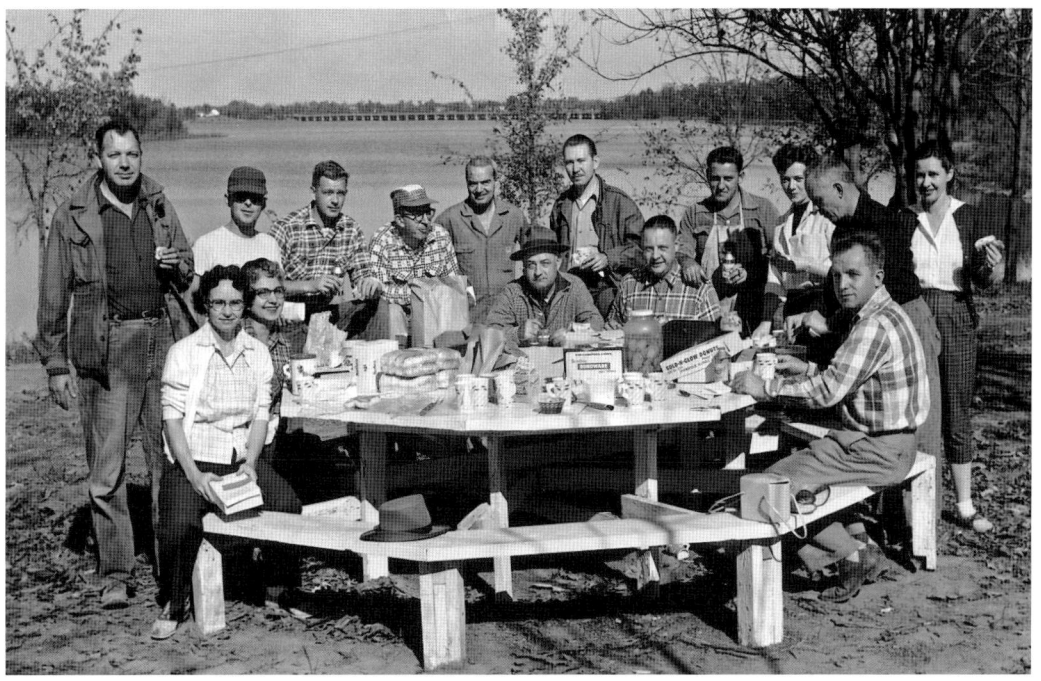

Good times at Bridgeview Beach, and during the construction of the TRN (Television, Radio, & Newspaper) Club, both at the western end of the lake. [*Sangamon Valley Collection, Lincoln Library*]

Lake Springfield in Illinois

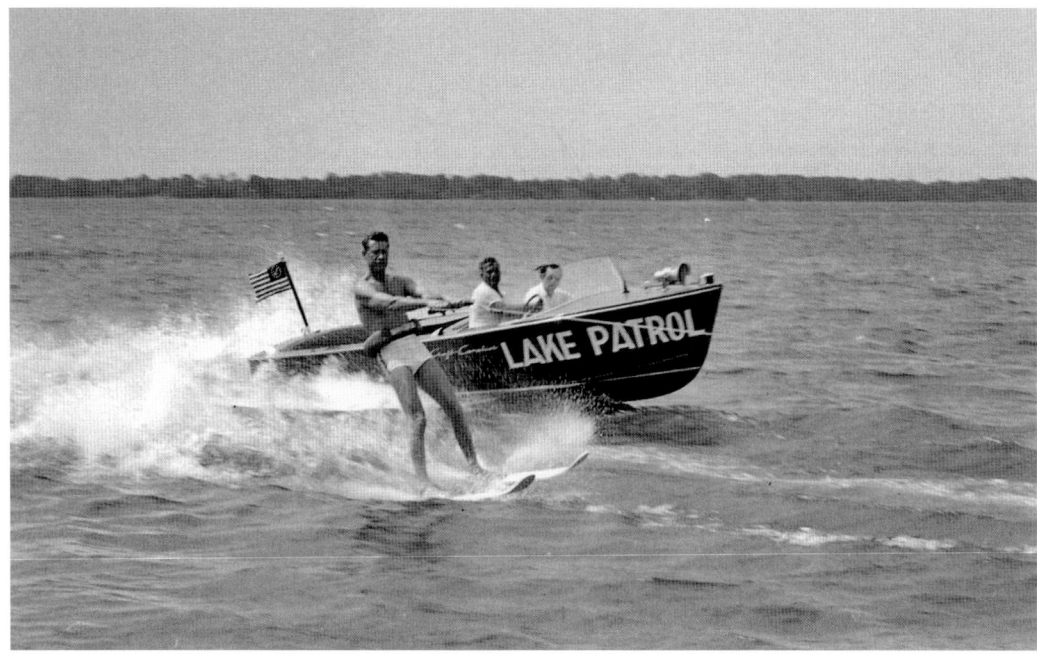

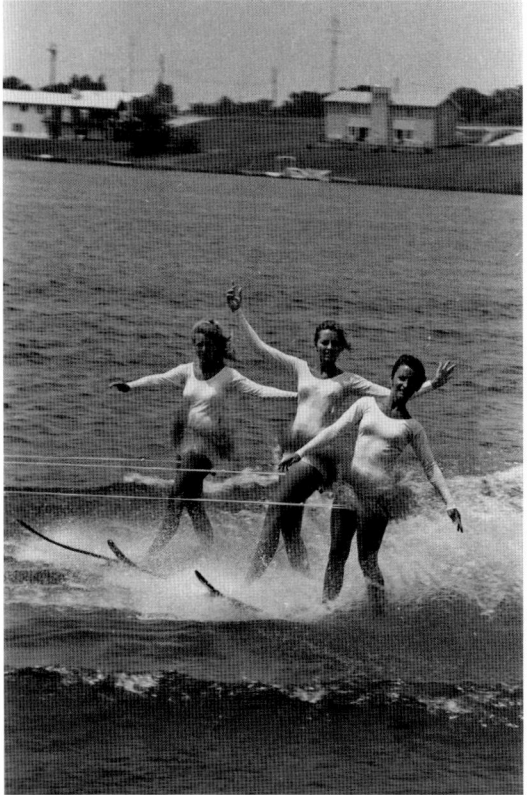

The Lake Patrol keeps a close eye on a skier, while trick skiers demonstrate their talents in front of the Ski and Boat Club. [*Sangamon Valley Collection, Lincoln Library*]

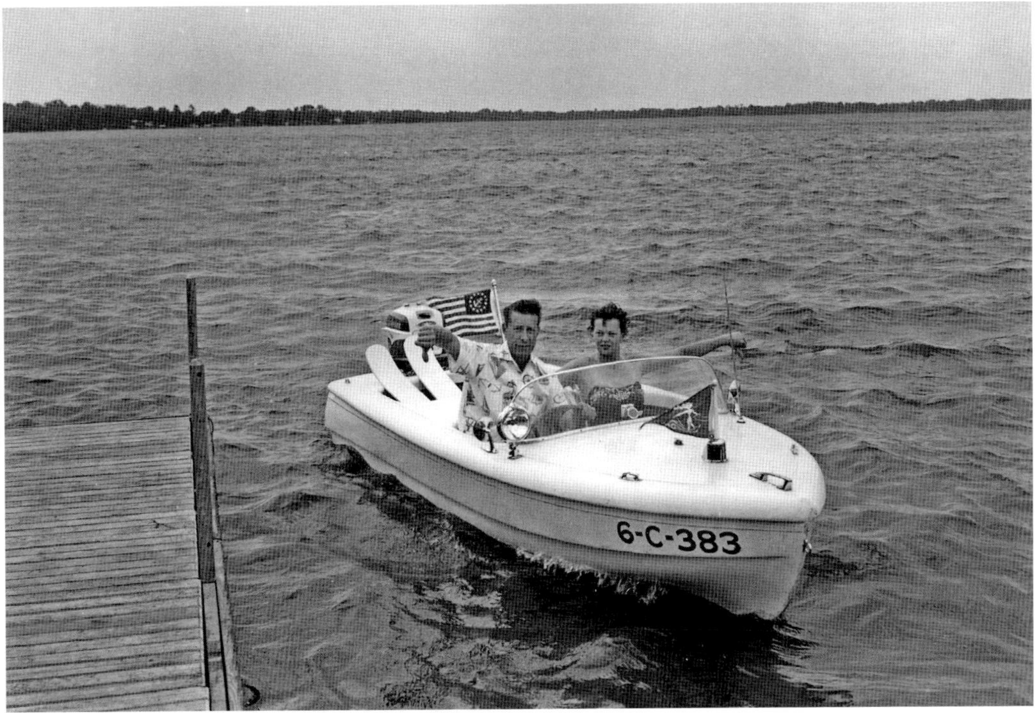

Life on the Lake

Boaters return to the dock after what was apparently a bad day on the lake (above), and a "sunfish" sailboat race near the Yacht Club. [*Sangamon Valley Collection, Lincoln Library*]

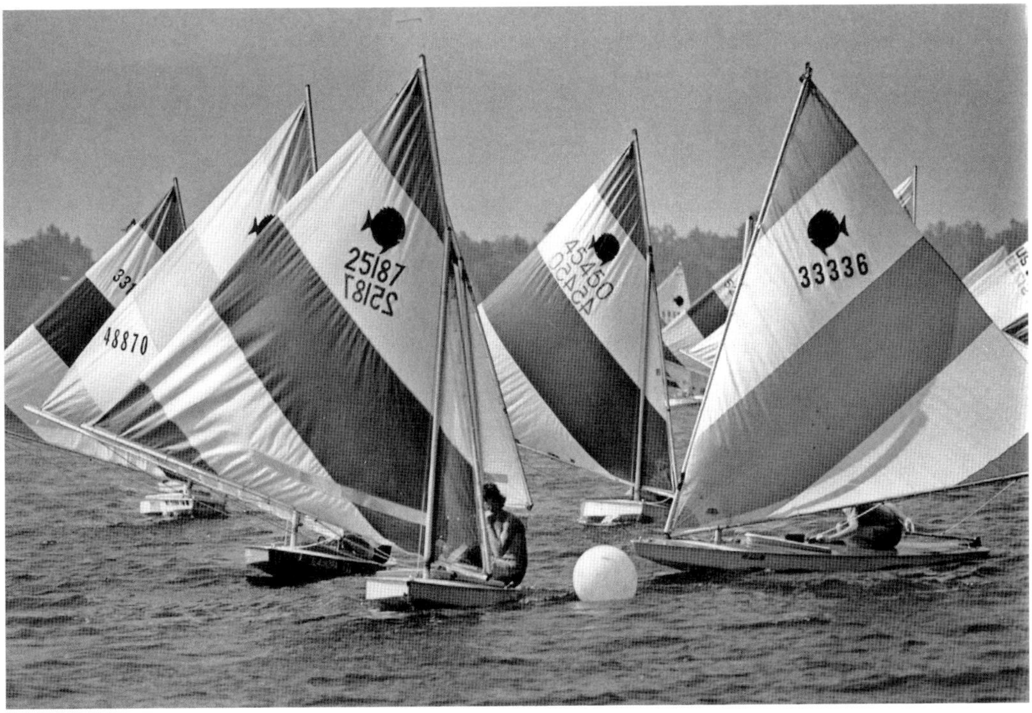

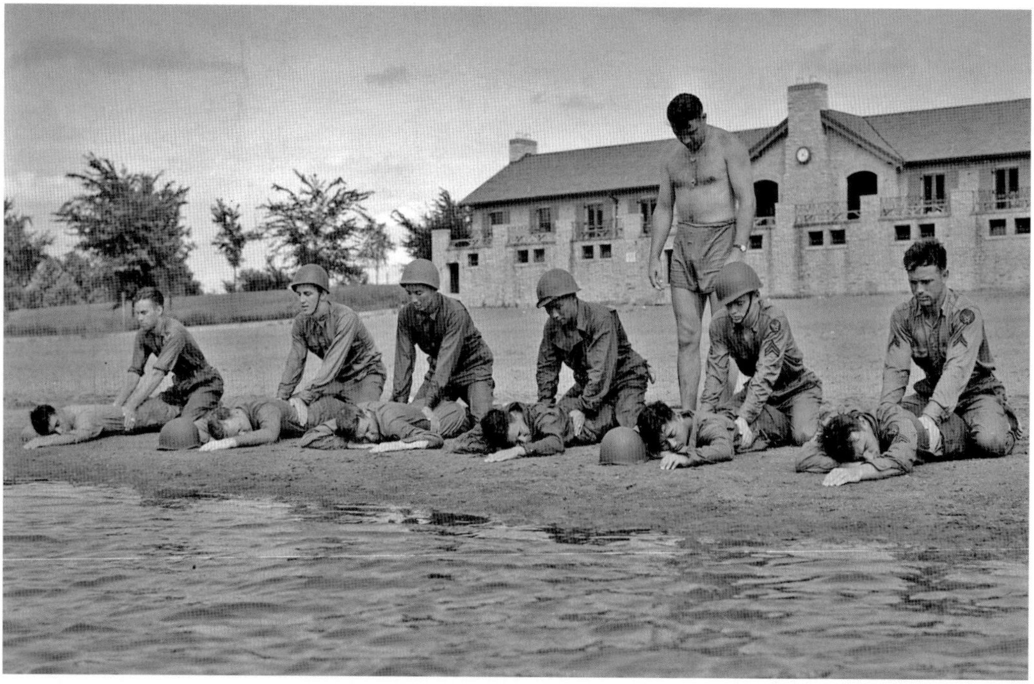

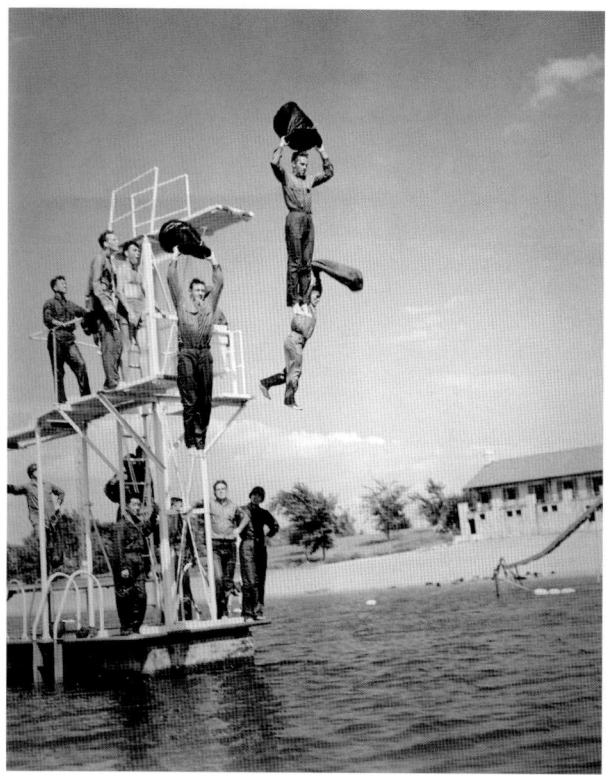

During World War II, soldiers practiced CPR and other emergency drills at the beach house. [*Sangamon Valley Collection, Lincoln Library*]

Life on the Lake

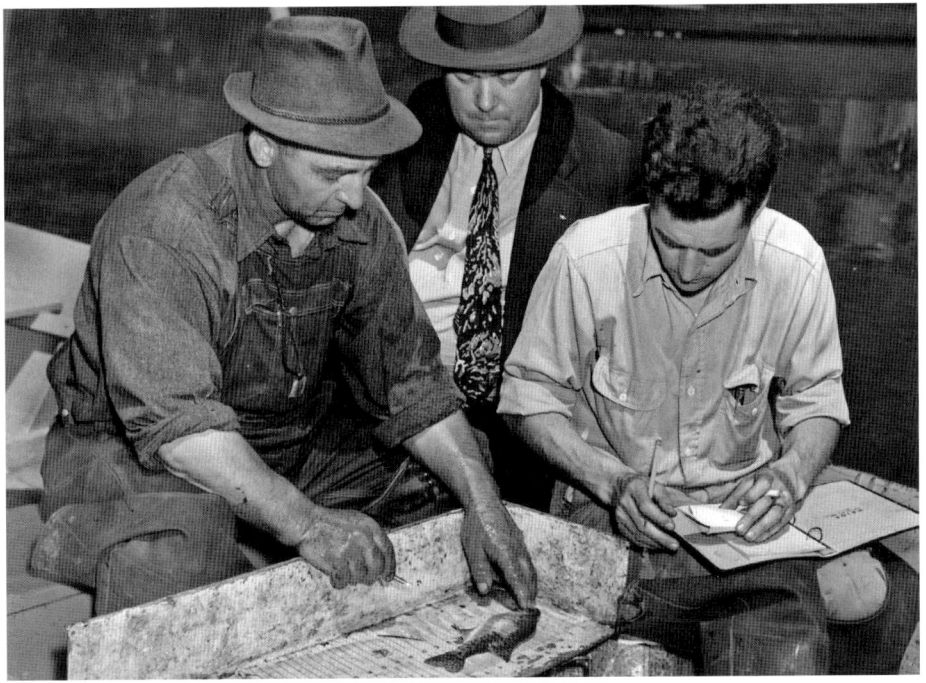

A study of the fish population of the lake in 1941 (above), and in the early spring of 1948, masses of dead shad appeared from beneath the thawing ice. [*Sangamon Valley Collection, Lincoln Library*]

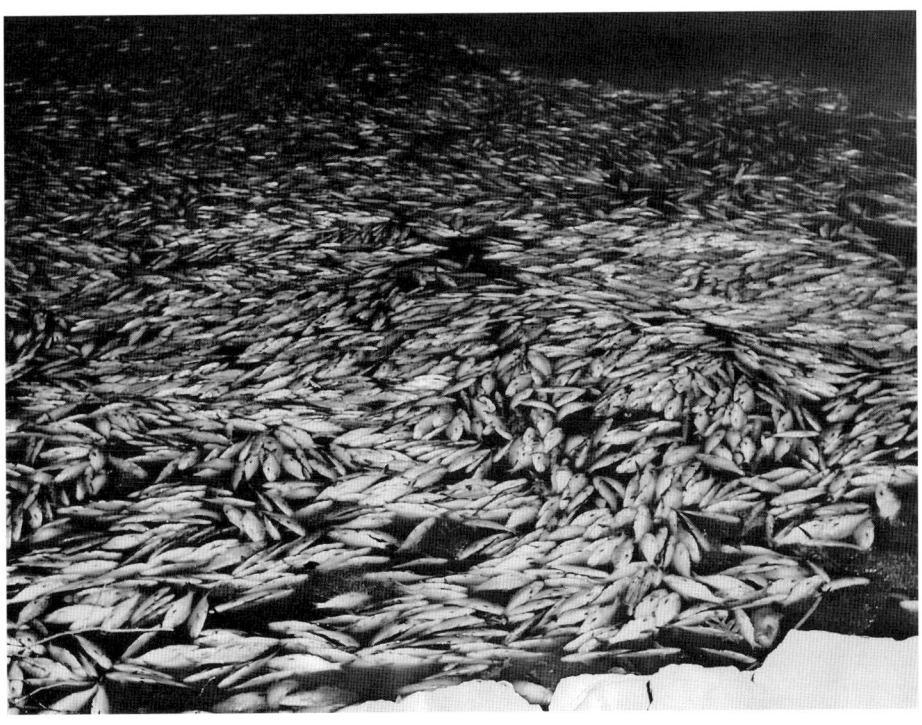

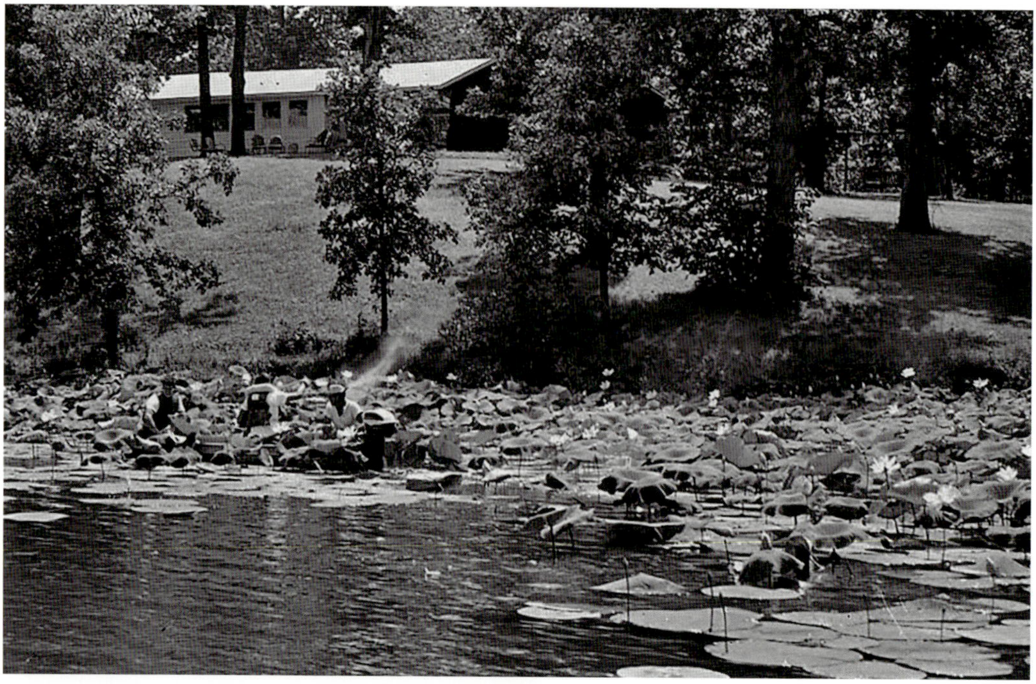

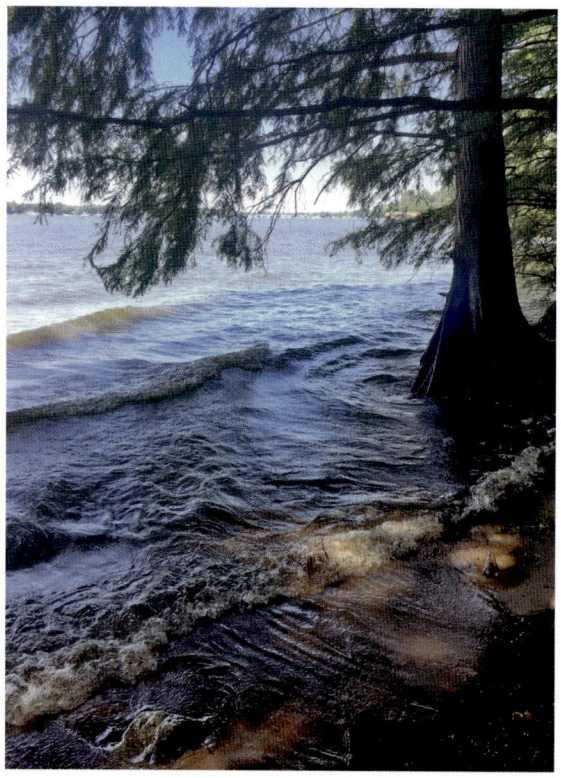

For much of the history of Lake Springfield, the coves and bays were filled with American lotus or "water lilies" in the summertime. The long, durable stems of these native water plants could easily choke the propeller of a small boat and made navigation difficult. Mechanical eradication programs of the 1970s destroyed most of the American lotus colonies in Lake Springfield. Below, a small cypress grove can be seen on the shores of Lincoln Memorial Gardens. [*Above: Sangamon Valley Collection, Lincoln Library*; *Below: Courtesy of Robert Mazrim*]

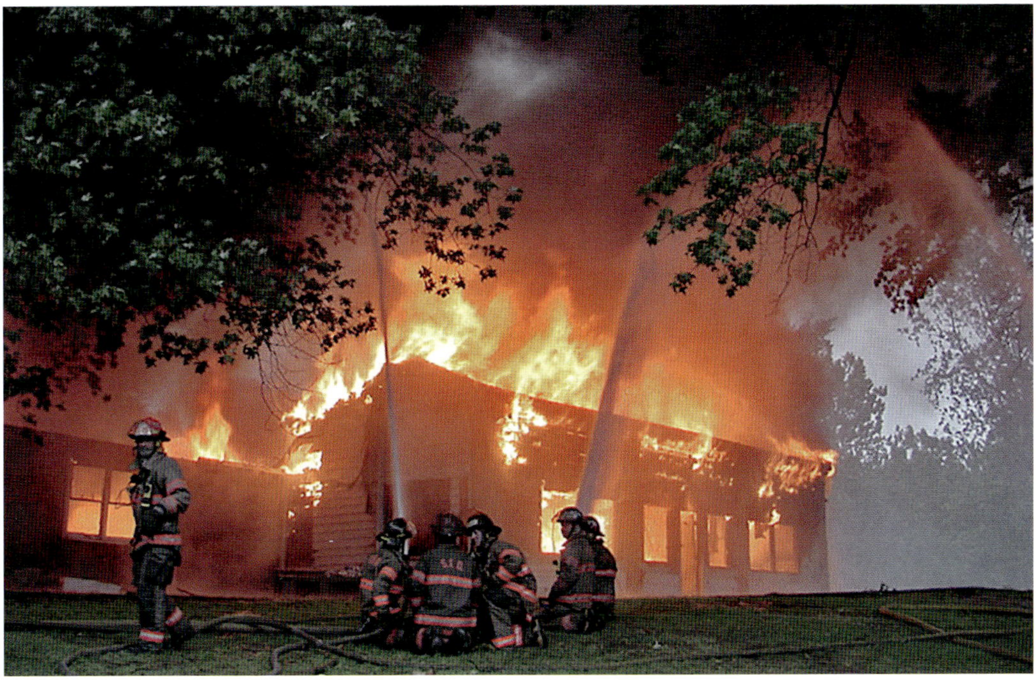

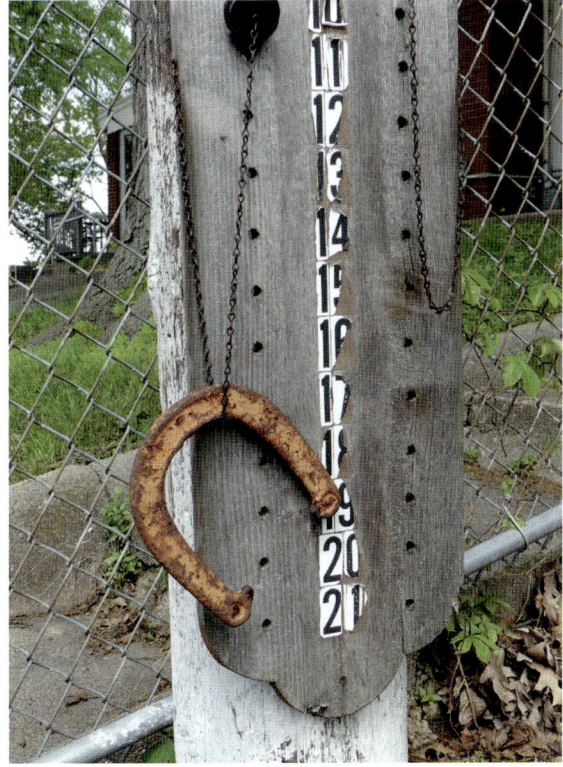

In August of 2002, the Springfield Fire Department conducted a controlled burn of the old Postal Club, in advance of constructing the new "Firefighters-Postal Lake Club." Below, a horseshoe game has fallen into disuse at one of the old lake clubs. [*Above: Courtesy of Firefighters Postal Club; Below: Courtesy of Robert Mazrim*]

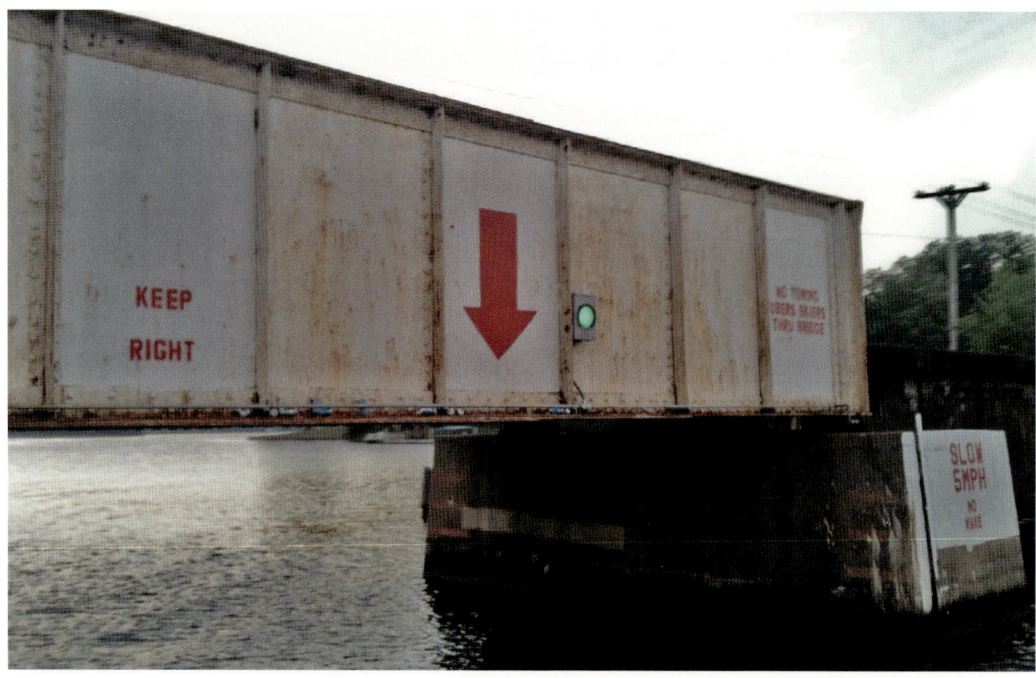

The old railroad bridge at Cotton Hill, seen at dusk in 2010, and the Spaulding Dam bridge at night, circa 1940. [*Above: Courtesy of Robert Mazrim; Below: Sangamon Valley Collection, Lincoln Library*]

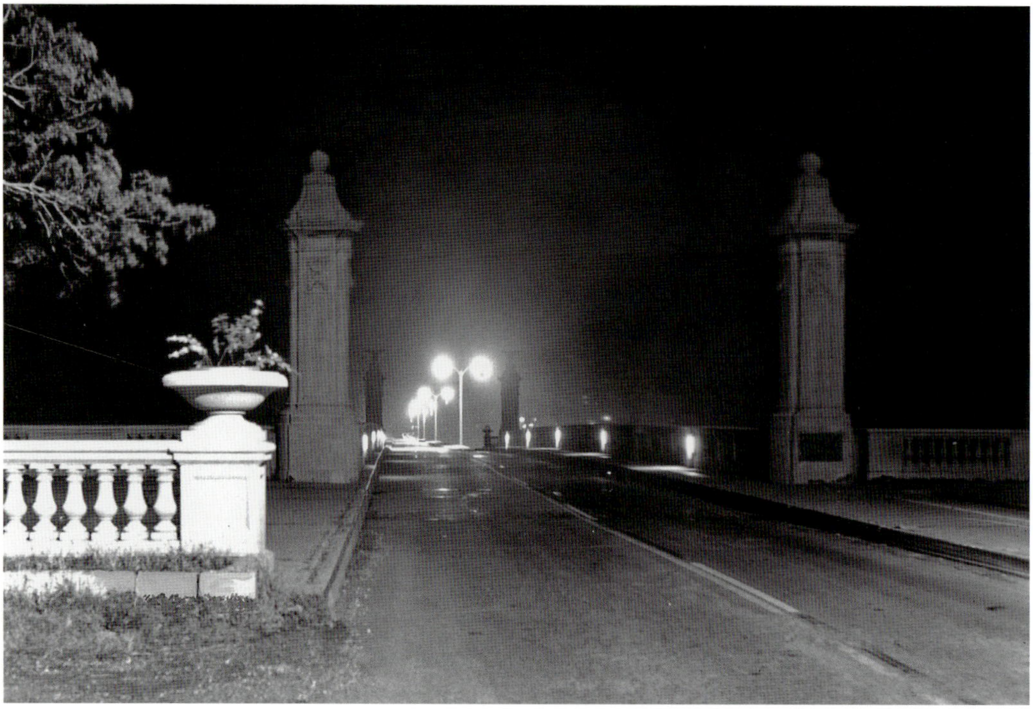

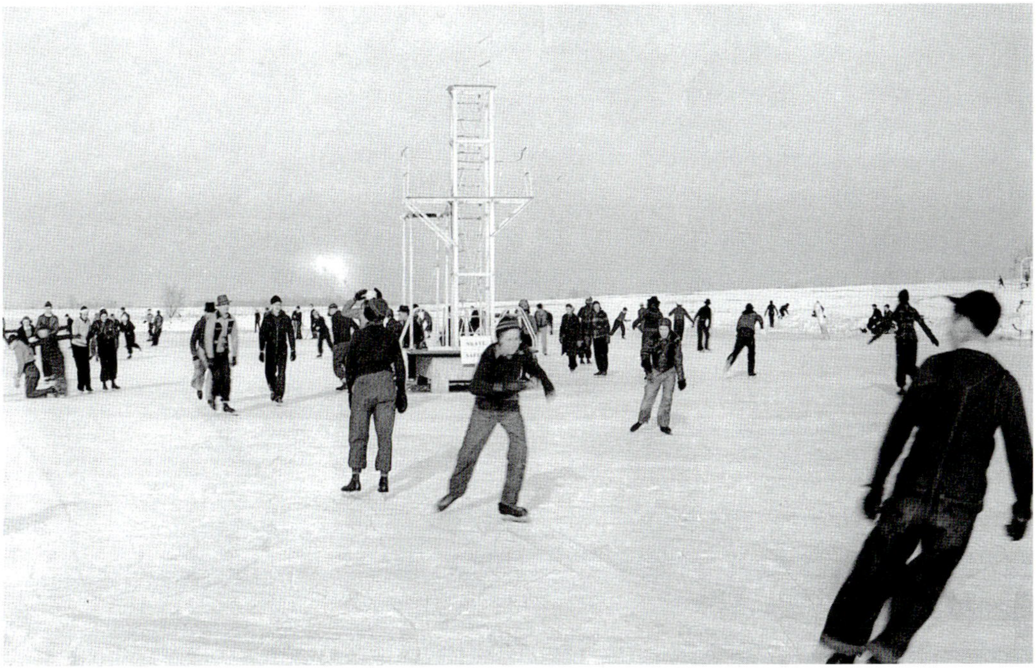

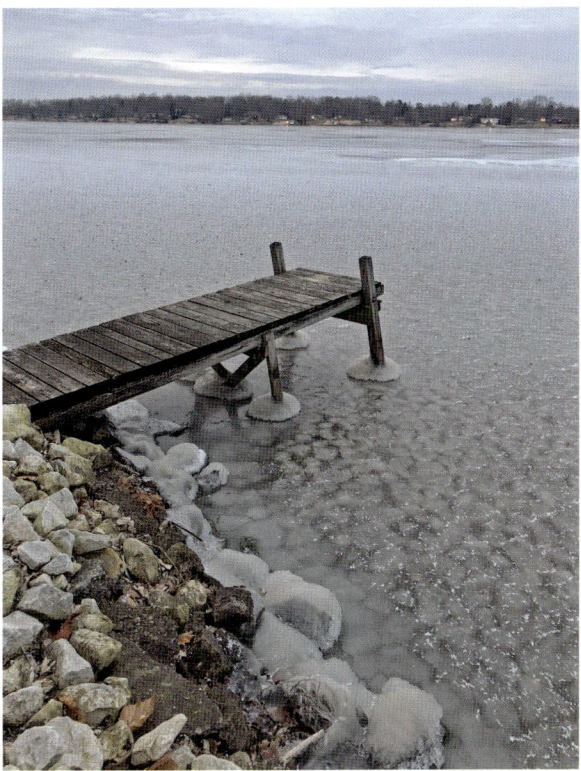

Ice-skating around the high dive at the Beach House around 1940, and an old fishing dock is locked in ice in 2019. [*Above: Sangamon Valley Collection, Lincoln Library; Below: Courtesy of Robert Mazrim*]

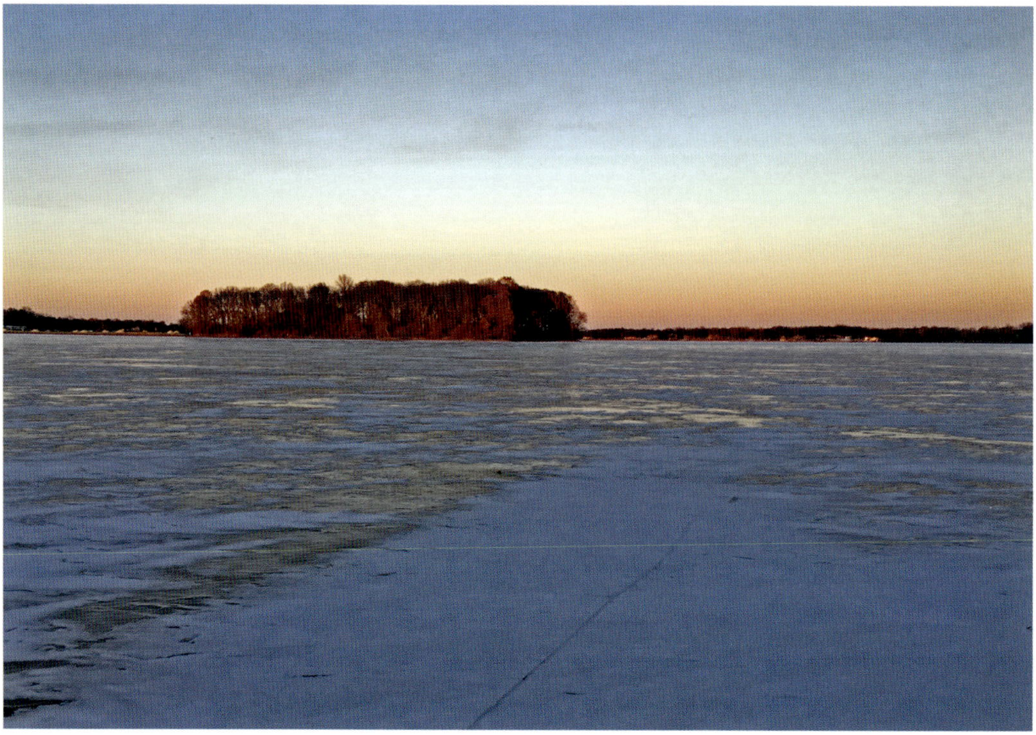

Lake Springfield in Illinois

A winter view of the frozen lake and Marine Island at dusk. Below, an icy winter shore. [*Courtesy of Robert Mazrim*]

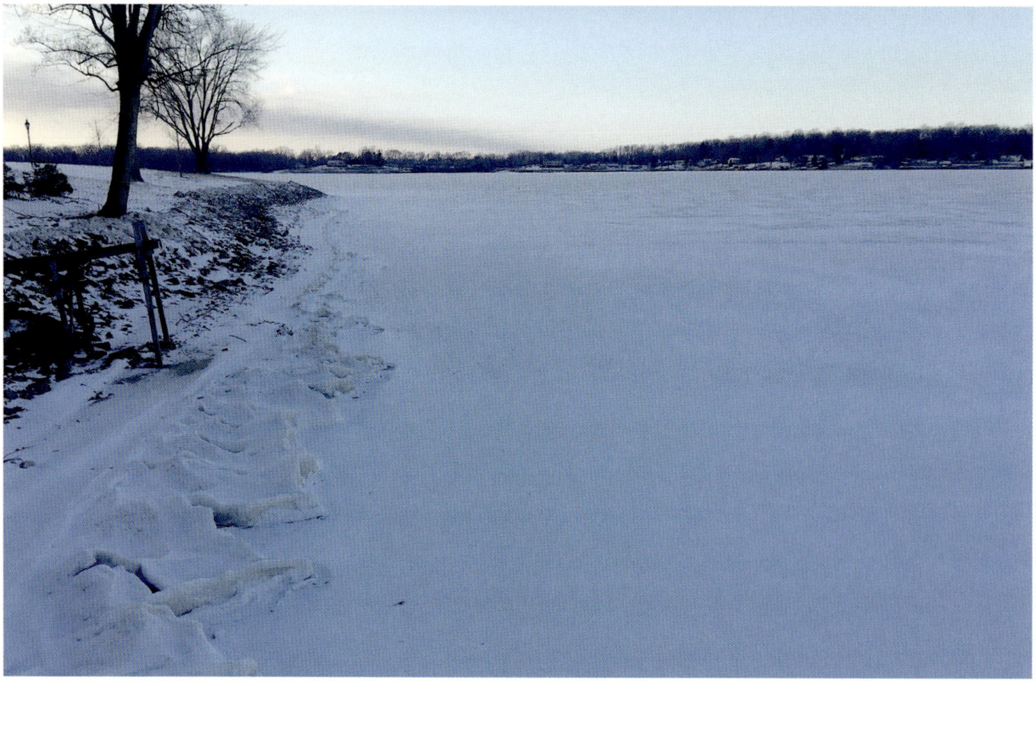

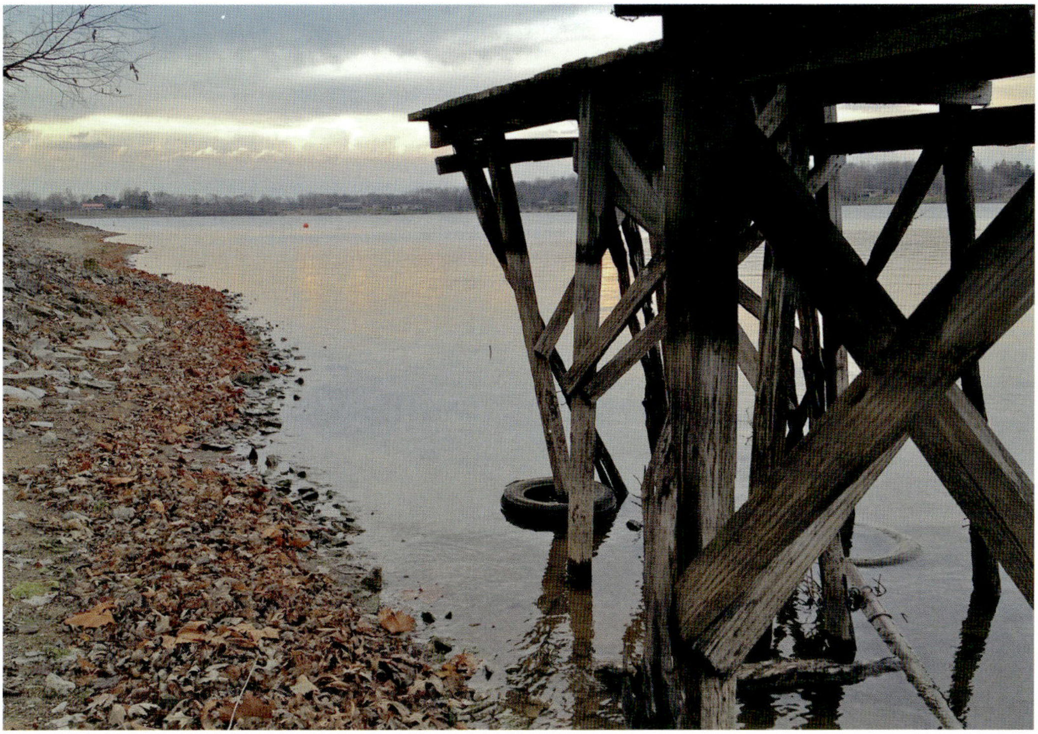

After years of occasional ice damage and general weathering, this homemade dock on Orchard Lane was deemed beyond repair. It was expected to crumble during demolition, but unexpectedly remained completely intact as it was lifted from the mud by a crane. Built of Osage orange (or "hedge") posts, the forty-plus-year-old dock was still largely impervious to decay below water. [*Courtesy of Robert Mazrim*]

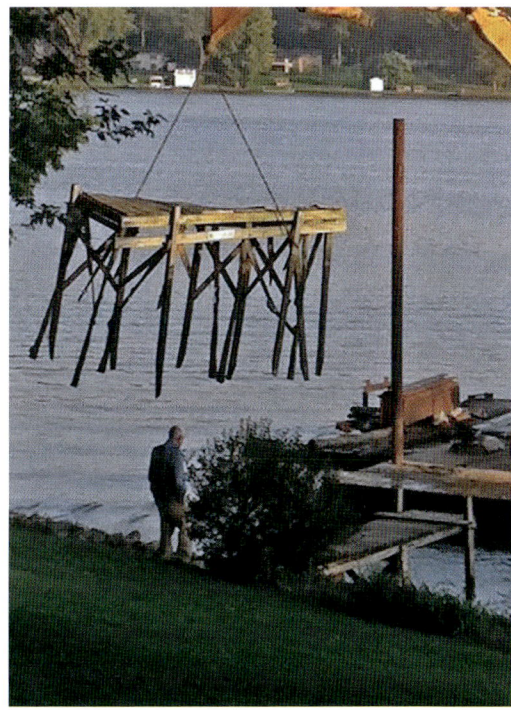

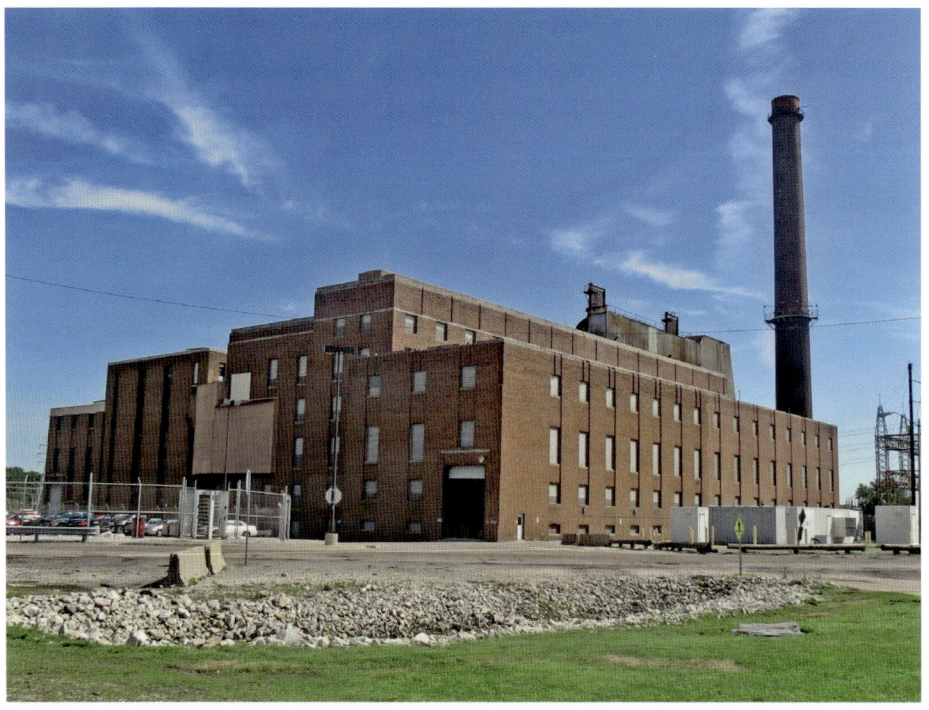

The original Lakeside power plant, as well as an early addition (above), now stands vacant. Meanwhile, the subsequent Dallman power plants, built in the 1960s and 1970s, are now facing closure. [*Courtesy of Robert Mazrim*]

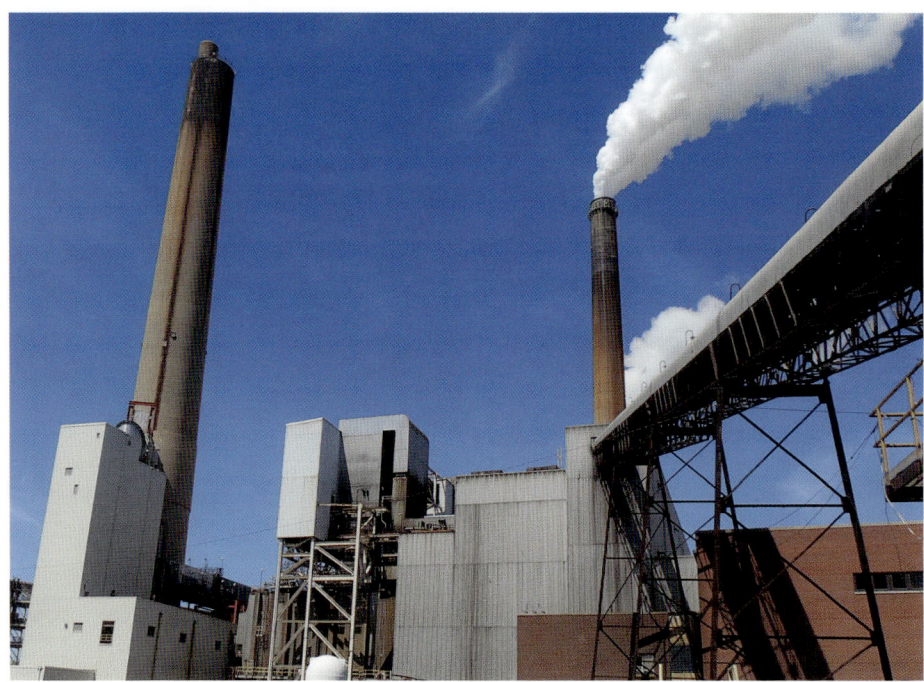

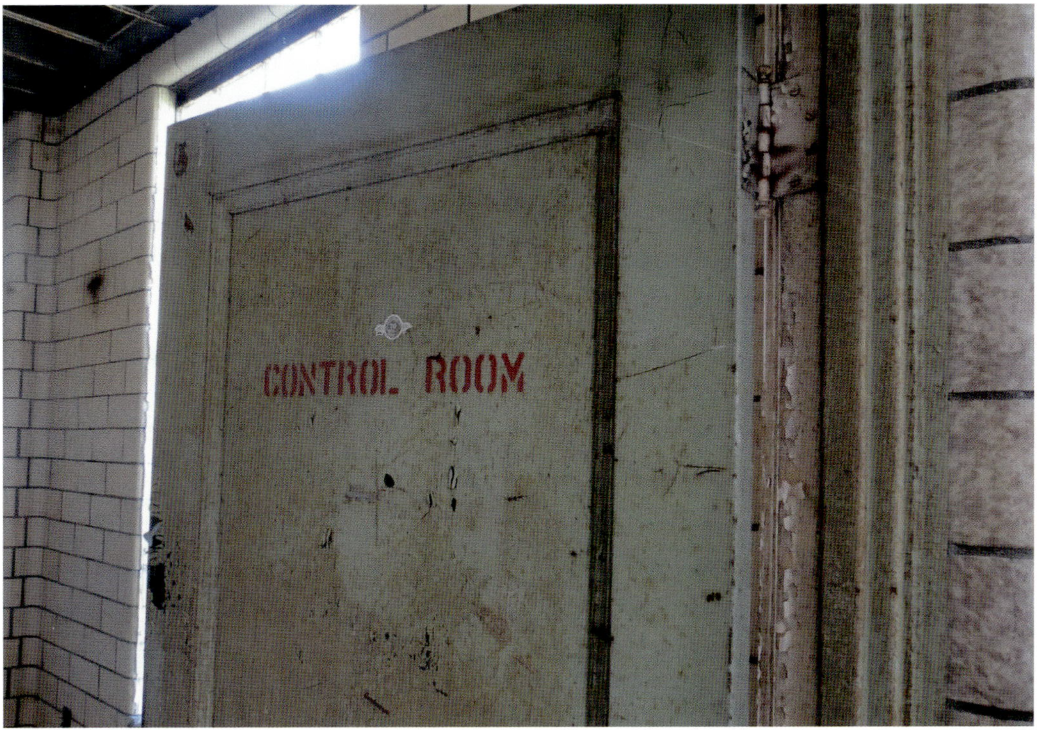

Inside the original Lakeside power plant, a door to an empty "control room," and gauges that no longer register electricity. [*Courtesy of Robert Mazrim*]

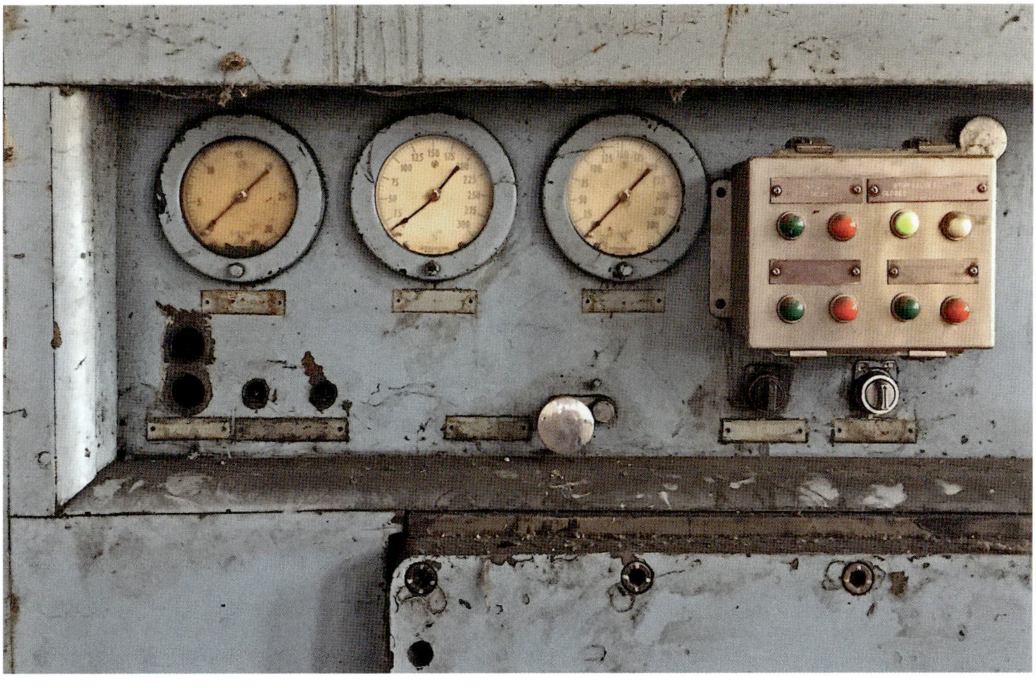

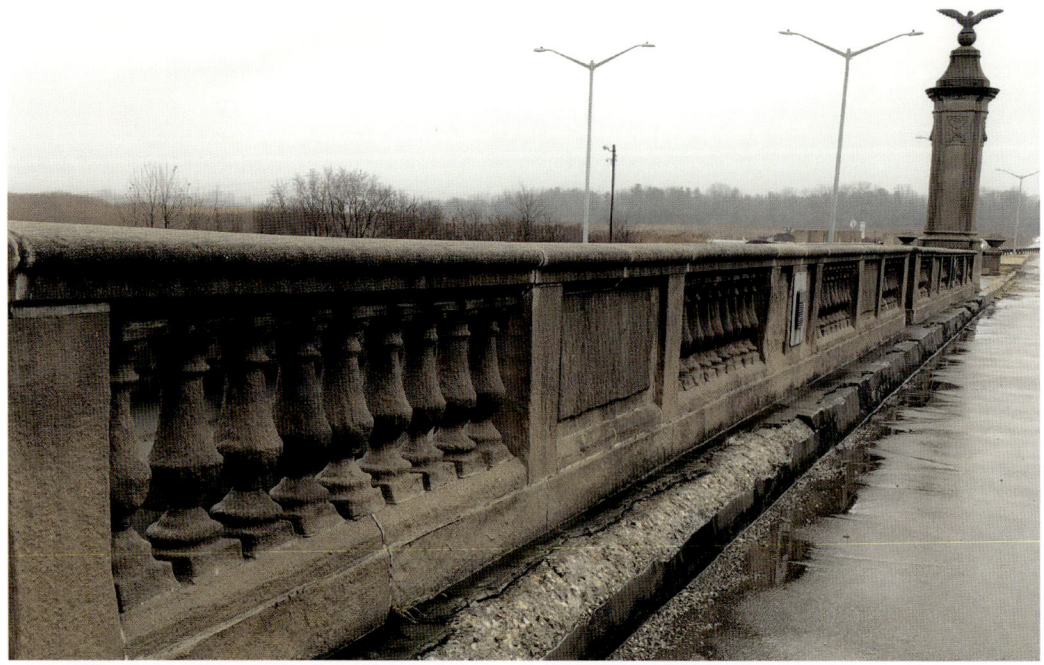

The 1930s concrete architecture on Lake Springfield is beginning to show its age in places. In 2011, Lindsay Bridge was extensively repaired and refurbished. Here, the concrete ornamentation on the Spaulding Dam Bridge shows the effects of over eighty Illinois winters, and needs repair. [*Courtesy of Robert Mazrim*]

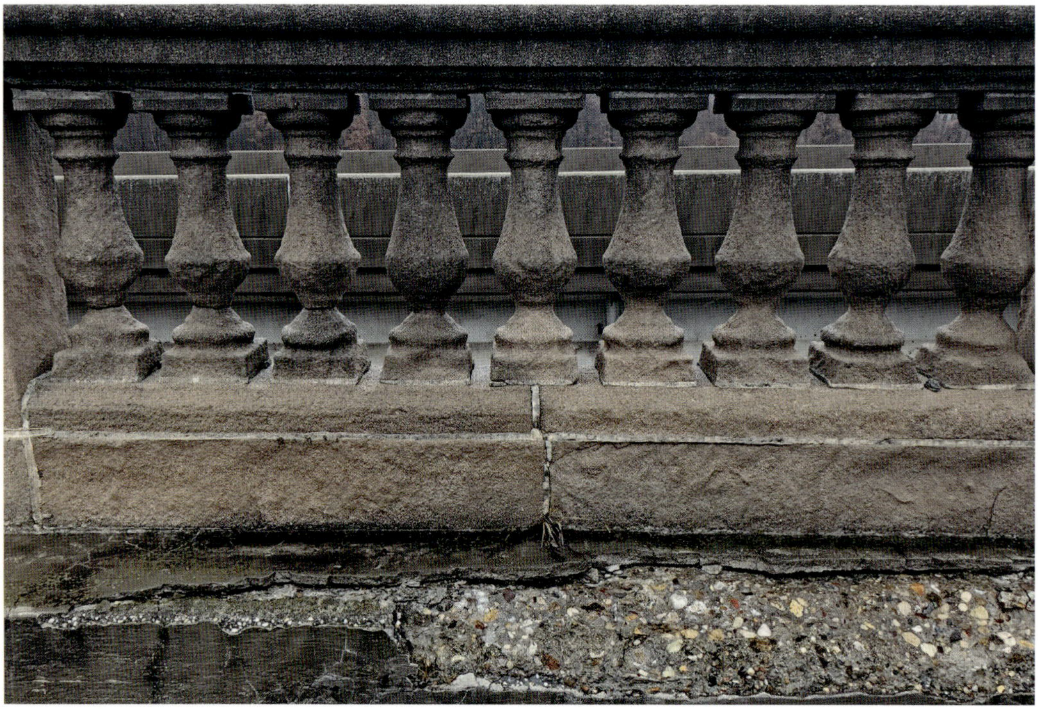

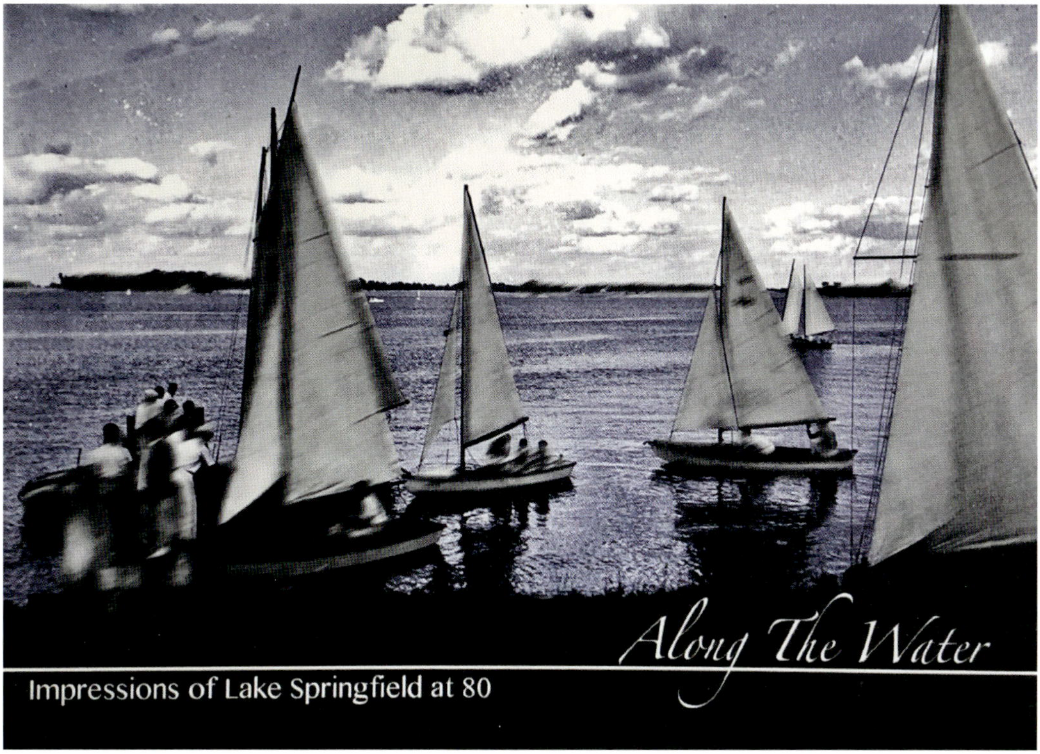

Life on the Lake

Along The Water
Impressions of Lake Springfield at 80

In 2015, the Springfield Art Association hosted an exhibit titled "Along the Water," which celebrated the eightieth anniversary of Lake Springfield, and featured lake-inspired works by a number of area artists. Below, "Bridge #1," a painting by Robert Mazrim and Erin Svenson. [*Courtesy of Robert Mazrim*]

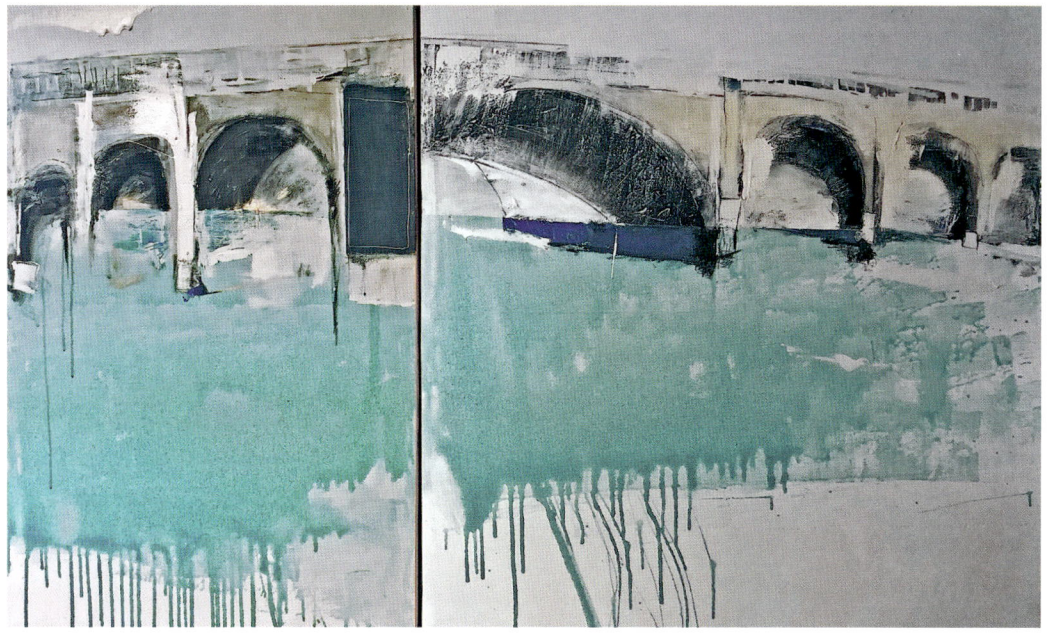

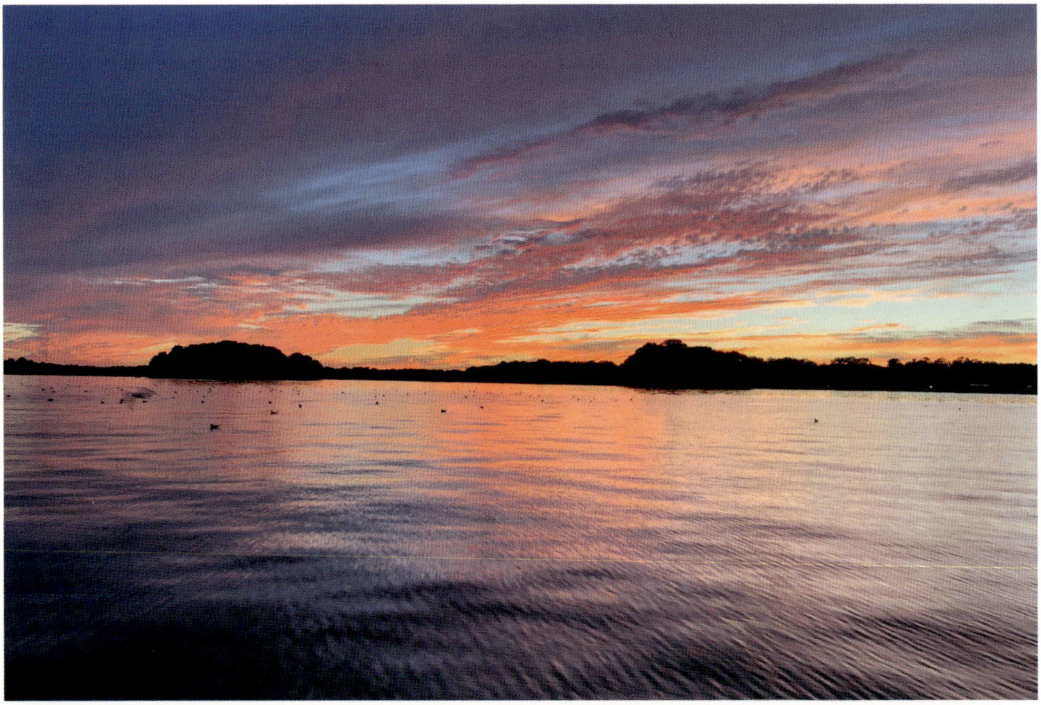

Lake Springfield in Illinois

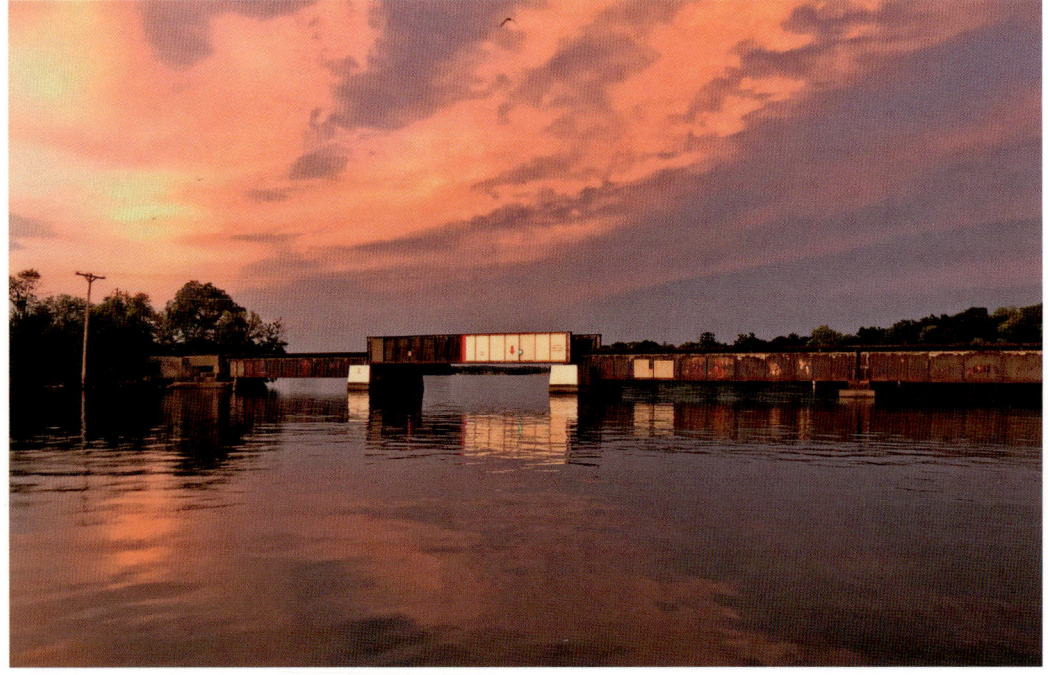

In September of 2020, the atmospheric effects of massive wildfires in California could be seen 2,000 miles away in the skies above Lake Springfield, casting the scene with an impressive red glow. [*Courtesy of Robert Mazrim*]